Paradox 10/10/93

PELICAN BOOKS

Pelican Library of Business and Management

TECHNOLOGICAL FORECASTING

Gordon Wills is Professor of Marketing Studies in the University of Bradford Management Centre, Chief Executive of Management Consultants (Bradford) Ltd, and a Director of Roles & Parker Ltd. He graduated from Reading University in Political Economy and read for the postgraduate Diploma in Management Studies, which he holds with Distinction, at night school in Slough. Prior to returning to academic life he held marketing research positions with ICI, Sales Research Services and Foote, Cone and Belding. He has lectured, broadcast and written extensively in Europe and North America, and recently spent six months as Visiting Professor of Business Operations at the University of Alberta. He is editor of the *International Journal of Physical Distribution* and the *European Journal of Marketing*. He has acted as consultant to many companies on management and marketing issues including Dunlop, ICI, IMI, UGI, Watney-Mann, Geigy (U.K.) and IPC. He was for three years an adviser on management topics to the BBC, and the Office for Scientific and Technical Information, and is currently a Board Member of the Centre for Physical Distribution Management.

TECHNOLOGICAL FORECASTING

The Art and its Managerial Implications

GORDON WILLS

in association with

RICHARD WILSON

NEIL MANNING

AND

ROGER HILDEBRANDT

PENGUIN BOOKS

Penguin Books Ltd, Harmondsworth, Middlesex, England
Penguin Books Inc., 7110 Ambassador Road, Baltimore, Maryland 21207, U.S.A.
Penguin Books Australia Ltd, Ringwood, Victoria, Australia

—

First published 1972

—

—

Made and printed in Great Britain by
Cox & Wyman Ltd, London, Reading and Fakenham
Set in Monotype Times

To my graduate and undergraduate students who miss my classroom appearances and tutorials whilst I take sabbatical leave, teach managers, research my subject, consult or write; and tell my superiors so without undue resentment.

Contents

Preface and Acknowledgements 11

Introduction *Technological Myopia* 15

Interaction of technology and the marketing concept; marketing myopia. Galbraith's Revised Sequence; revising that sequence. The Educational and Scientific Estate. The social planning of technology. Technological forecasting in the corporate context.

PART ONE THE MANAGERIAL IMPLICATIONS OF TECHNOLOGICAL FORECASTING

Chapter 1.1 *The Technological Setting* 37

Forecasting as an eternal problem. Need for integrated consideration of economic, social and technological elements. Lengthening time-scale of investment. Continuous innovation a contemporary phenomenon. Present decisions; future values. Imagination tempered by extrapolation; images of the willed future. The efficacy of technological forecasting and its limitations. Glimpses of the attainable future.

Chapter 1.2 *The Environmental Impact* 48

The environmental factors. The relevance of social objectives; success as a goal and social sacrifice. Objectives at the corporate level; the technostructure; the individual and the organization. Goal-striving behaviour. Integration of social and technological change; exploration of interrelationships. The role of government to give social priorities in the industrial system; the aesthetic dimension. Internationalization of technology.

Chapter 1.3 *The Corporate Impact* 64

Sensitive management variables. TF and the corporate planning function; integration. Concept of coupling. Weaknesses in strategic planning; alternatives. TF and the planning process – policy, strategic and tactical levels. The forecasting cycle. Organizing to meet forecast technology; structures; function orientation. Diffusion rates and their implications. Technological mapping.

Chapter 1.4 *The Functional Impact Within Business* 79

International markets. Shorter diffusion and life cycles. Resistance to innovation; marketing's educational role. Transportation developments opening wider markets. Financial management. Risk reduction; cost–benefit analysis. Research and development; fundamental planning cycle and offensive/defensive research. Utilization – state of the art/body of knowledge; relevance analysis and criticality of component technologies; deficiencies; research programmes. The innovative role of the small firm. Productive aspects; design; raw materials; productivity. Personnel aspects; unskilled work force. Training and retraining. Inputs for coherent manpower planning and organizational development. TF as a tool for coping with the process of technological change.

PART TWO THE FORECASTING ART

Chapter 2.1 *Extrapolative Approaches* 106

Functional capabilities; specific technologies; scientific and technical findings as data inputs. Four classes of curve; linear, S, and exponential; examples from mining productivity, energy conversion, efficiency for illumination, lasers, speed of man. Constraints, intensive and extensive; time and parameter constraints; examples from hovercraft, computing speeds and desalination. Envelope curves. Technology precursors; multiple trends. Floyd's phenomenological model for trend extrapolation; his nomograph. Learning curves and budgetary allocations. *Appendix*: Estimation of confidence intervals in a forecast.

Chapter 2.2 *Morphological Analysis* 131

Classical origins with Aristotle: Lull's attributes of God; Goethe and Darwin; Reuleaux and the Kinematics of Machinery. Zwickey's rules for definition and derivation of parameters; meaningful alternative combinations. Matrices from marine communications, functional fluids, retail sales data capture, commercial dating, ocean transport, metal working. Screening of alternatives; small perturbation approach. Overlapping systems and process analysis – design and beer production. Morphological mapping in hyperspace; breakthrough opportunity index. Matching with market needs; gap analysis and system sensitivity.

Preface and Acknowledgements

THIS book is written for everyone who is concerned with the management of technological change, whether as agents for such change, or as recipients of the benefits and discomforts it brings in its wake. In Part One I have tackled the managerial implications for both the total environment and for the firm, as a prelude to identifying and describing the techniques which are most commonly used, but yet only recently developed in a formalized way.

Technological forecasting is not a grand new concept; rather it is an assemblage of techniques which have gradually emerged in a variety of countries to coalesce into a powerful new technology for the mastery of technology itself. There is little doubt that although knowledge of the techniques and potential for technological forecasting is not particularly widespread, it is a subject of which a great deal more will be heard in the seventies. It had its origins, as did operational research and industrial psychology before it, in urgent military needs. Whilst industrial psychology advanced dramatically during the First World War as a basis for selection procedures, and OR to fight the air and sea battles of the Second World War, technological forecasting and the networking procedures emerged in the 1950s and 1960s to aid the missile arms race and space research. It has until recently been the exclusive tool of companies and agencies working in advanced technology sectors, most noticeably on defence contracts. Hence I firmly hope that this book will be of value to a wide range of industrial practitioners in research and development, marketing research, financial management and corporate planning. The evidence I have collected during the past five years points strongly to the value of technological forecasting as a tool for the detailed analysis of research and development budgeting procedures, and this of course has dramatic significance for corporate planners.

My involvement in the study and application of technological forecasting has been an object lesson for me in the marketing of fashion goods. The exceptional enthusiasm which has characterized the business search to familiarize itself with this area has led me to put pen to paper and prepare this book. In the course of five years, largely through the medium of our Post-experience Programmes at Bradford Management Centre, I have had the opportunity to meet and discuss technological forecasting with most of the world leaders in this subject and most of the European firms engaged in its application. In particular, I have been privileged to work with Dr Jantsch, from the OECD; Professor Brian Quinn, from the Amos Tuck Graduate School of Business; Professor James Bright of Harvard and now Texas University Graduate School of Business; James Hetrick, Senior Consultant with Arthur D. Little; Dr Igor Ansoff, of Carnegie-Mellon University, now Dean of Vanderbilt Graduate School of Business, Tennessee; Harry Jones and Dr Grigor from Geigy (UK) Ltd; Peter Hall from ICL; and John Dingwall from NCR Ltd both in Europe and at Dayton, Ohio. From each of these, and from many more, I have received encouragement and inspiration for the translation of the jargon of technological forecasting into a comprehensible, manageable form, which will facilitate its diffusion and widespread adoption where necessary in industry.

I myself am no technological forecaster. To do that one needs extensive technological knowledge of the field in question. I hope, however, I have grasped its essence and its implications and that in this book I can pass them on.

In coming to my own personal understanding of technological forecasting and its implications I have been enormously helped by discussions with managers attending our conferences and seminars at Bradford University, and at in-company programmes with the then Ministry of Technology, NCR, and Geigy. Many fruitful scenario-writing seminars and morphological analyses have taken place on these occasions and have extended my understanding of how these techniques in particular can be deployed. Equally supportive illustrations of S and envelope curves for extrapolative forecasts have been plentifully provided. My thinking in this direction was given its major fillip, however, by the *First National Conference on Technological Forecasting* which David Ashton,

Bernard Taylor and I organized at Bradford in association with *The Times* and the then Ministry of Technology in July 1968. From Mintech the stimulus of the Programmes Analysis Unit at Didcot through Ken Binning and Tim Garret, and of their Leeds director, Keith Duxberry, was especially valuable. The proceedings of that conference were subsequently published by Crosby Lockwood for Bradford University Press under the title *Technological Forecasting and Corporate Strategy* (1969); in North America by American Elsevier; and in Japan.

In the preparation of this particular book, I have worked in close association with Richard Wilson, Neil Manning and Roger Hildebrandt. Richard Wilson was from Autumn 1969 to Summer 1970 Roles and Parker Research Associate at the University of Bradford Management Centre and he prepared, *inter alia*, early drafts of Part One and the Glossary of Terms. Neil Manning and Roger Hildebrandt were my Research Assistants at the University of Alberta during Spring 1970 and prepared rough drafts of Chapters 2.1, 2.2, 2.4 and 2.5. (My Bradford colleague, Martin Christopher, helped crystallize my views on Mission Analysis which are discussed in Chapter 2.5.) For these several assistants I have H. T. Parker, Chairman of Roles and Parker Ltd, to thank and Dean Ted Chambers of the Faculty of Business Administration at the University of Alberta. Dean Chambers also permitted me the time during my sabbatical leave of absence in Canada to write most of the final draft of this book and afforded me the facilities to turn it into an acceptable typescript.

Finally, I wish to thank the editors and publishers of *Management Decision*, *Journal of Long Range Planning* and *Commentary*, *Journal of the Market Research Society*, for permission to reproduce in part some of my own material which originally appeared on their pages. Full details of the articles from which I quote are given in the References to the Introduction cited at the end of this book.

Management Centre
University of Bradford

Gordon Wills
June 1970

INTRODUCTION

Technological Myopia

MY interest as a marketing executive, and later as an academic, in the industrial problems inherent in contemporary technological change was aroused by J. K. Galbraith. In this I am by no means unique. It has led me over the past five years to an extended examination of the processes of technological change and in this book I attempt to recount what I found. Most significantly I came across an emerging body of techniques called 'technological forecasting' which promised to offer much greater potential for the positive planning, coordination and control of technological change in the future. These techniques, and, more importantly, their managerial implications, form the substantial body of this book. At the outset, however, I wish to question some of the more popularly accepted ideas of Galbraith – not because I do not hold him in the greatest respect, but because I feel he frequently does less than justice to brilliant ideas by alienating them from many typical industrial situations.

It is my purpose here to explore the ways in which the imperatives of modern technology interact with the marketing concept; with the idea that the customer is in some sense sovereign in the determination of what will be produced and how it will be marketed in our economic system.

I have selected as the title to this Introduction a phrase which is intended to be quizzical and provocative. It is intended to jar with the over-simplification implicit in *le client-roi.* Technological myopia is a widespread disease amongst the business community, not least among its marketing members. It is a condition which requires careful handling, for much is at stake. But it is probably an unavoidable and necessary counter-condition to corporate attempts in recent years to scan the distant horizon to avoid charges of *marketing* myopia.

Marketing's recent ascendancy has meant a series of corporate adjustments to the customer's needs and wants. Maturity, however, for the marketing-oriented business can only be manifest in a proper relationship between the customer's interest and the intelligent use of research and development. One particular adolescent belief was that marketing's role was to plan and coordinate the development of the company's future unaided and independently of other functional interests. Such a viewpoint reaches its most subtle if extravagant exposition in Leavitt's much-discussed views on marketing myopia. His strictures, for instance, on the buggy whip and the petroleum industries are now enshrined in our mythology. Yet they need to be desecrated if we are to arrive at any currently valuable working relationship between these two, mutually dependent, founts of new product ideas. No company is a slave to any definition of the business it is in, nor even to its current pattern of resource allocation. The whole rationale of the corporate planning backlash towards marketing is to demonstrate that a company has a wide range of resources, each, some, all, or none of which can form a basis for successful growth, maintained performance, or whatever other objective we might pursue. Successful innovation, however, will normally require both marketing and technological innovation, and marketing and technological critique. The two are so closely interrelated that there cannot be work in one area without assumption and consequence that bear on the other. Product/market characteristics mutually determine one another. Schon has discerned that much of the problem stems from the professionalism of the division of labour between marketing and technology. Each tries to make true, safe statements within its professional territory, leaving the uncertainty inherent in the situation to the other. The problems of institutionalization once again rear their heads; once again we are faced with the need for a dynamic sociology if the problem is to be conquered. The problem has been described as sclerotic and is one of the major challenges facing the social sciences today.

GALBRAITH'S REVISED SEQUENCE

To meet the above challenge something greater than Galbraith's 'revised sequence', which he develops fully in *The New Industrial*

State, is required. The synergy which can result from the interaction of technology and marketing is not a sinister social conspiracy, as Galbraith is tempted to imply, by a small but powerful group of technologists, dubbed the technostructure, but a more satisfactory, less wasteful deployment of business resources in the service of the customer and hence of society.

Galbraith's thesis is worth restating here to demonstrate his synergetic myopia:

> We have seen that the accepted sequence [i.e. customer sovereignty] does not hold. And we have now isolated a formidable apparatus of method and motivation causing its reversal. The mature corporation has readily at hand the means for controlling the prices at which it sells as well as those at which it buys. Similarly, it has means for managing what the consumer buys at the prices which it controls. This control and management is required by its planning. The planning proceeds from use of technology and capital, the commitment of time that these require, and the diminished effectiveness of the market for specialized technical products and skills.
>
> Supporting this changed sequence is the motivation of the technostructure. Members seek to adapt the goals of the corporation more closely to their own; by extension the corporation seeks to adapt social attitudes and goals to those of the members of its technostructure. So social belief originates at least in part with the producer. Thus the accommodation of the market behaviour of the individual, as well as of social attitudes in general, to the needs of producers and the goals of the technostructure is an inherent feature of the system. It becomes increasingly important with the growth of the industrial system.
>
> It follows that the accepted sequence is no longer a description of the reality and is becoming ever less so. Instead, the producing firm reaches forward to control its markets and on beyond to manage the market behaviour and shape the social attitudes of those, ostensibly, that it serves. For this we also need a name and it may appropriately be called The Revised Sequence. I do not suggest that the revised sequence has replaced the accepted sequence. Within the industrial system the accepted sequence is of diminished importance in relation to the revised sequence.

Galbraith does, of course, go on to agree that the two can exist side by side, and I, for one, have no wish to deprecate the direction of his argument. None the less, he fails to explore the significance of the two-directional sequence. Some aspects of demand satisfaction do indeed require a long planning period with concomitant high risks, and a need to manage demand if at all possible. But there are, equally, a wide range of technological improvements and innovations which do not. Furthermore, the corporate need for managed demand (from the firm's viewpoint) can arise from planned marketing investment as well as from planned industrial investment. Marketing investment can often be substantially higher than its manufacturing equivalent.

That a technostructure exists in industry is beyond doubt, and the reasons Galbraith ennunciates for its emergence are surely right. Most industries today employ the widest possible range of knowledge to meet the complex technological demands that are made upon them. This knowledge is shared amongst many individuals and must be integrated. In addition, the forward commitment implicit in the gestation of new technology makes the forecasting and/or creation of specific demand fundamental to corporate success. Team reconciliation is accordingly unavoidable, but Galbraith again strays when he suggests this is exclusively the task of the technostructure. There are at least two readily identifiable counter-trends which are restraining the entrepreneurial/business-policy function in business. The first has already been alluded to as the corporate planning backlash. The corporate planning function provides top management with its own specialist skills in the integration of technologies and functions – it could perhaps be termed a techno-superstructure, but it is scarcely ever allied to the technostructure which continues to exist, but at a lower level. A second more problematic development is in terms of the increased potential for entrepreneurial decision-making with the advance of computing power. It is apparently now beyond all reasonable doubt that we shall, within the next two generations of computers, be able to restore our individual mastery of complex organizations to at least that level formerly enjoyed by the dominant figures of the nineteenth century.

None of these comments is intended in any sense to contradict the self-evident fact of which the marketeer is aware; that specific demand can be managed in much the same manner as the level of

Exhibit 1. *Institutional Promotion to Manage the Level of Aggregate Raisin Demand*

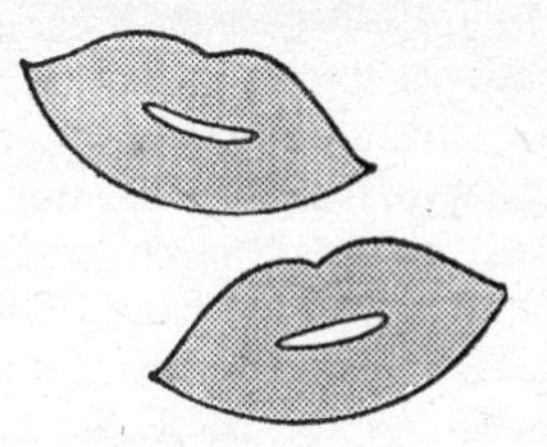

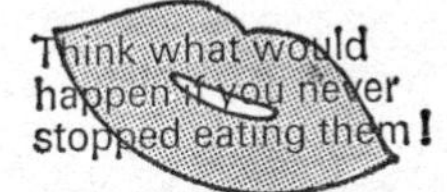

You can kiss
your husband 11 times
on the energy you
get from 12 raisins.
Think what would
happen if you never
stopped eating them!
And for fun new ways of eating these sweet little morsels of instant energy, send for our FREE new recipe booklet, 'Raisin Potpourri'. It's just filled with tempting, delicious raisin recipes. Try them. You may not kiss anyone, but you'll sure have a gleam in your eye from all that raisin energy!
To: Raisin Recipes
Clinter Station, Fresno, Calif. 93703
Please send my FREE 'Raisin Potpourri' recipe booklet.
Name
Street
City
State
Zip
TV

Source: TV Guide for Western Canada, 7 March 1970.

aggregate demand is managed within the total economy, i.e., along Keynesian lines. Institutional advertising in many sectors, of which the most delightful example I have ever seen is shown as Exhibit 1, has borne witness to this; so has the acceleration of consumption in response to educational campaigns during the early stages of product diffusion. Equally, however, we are aware that, unless the product does strike a chord, the sustained level of specific demand will not be maintained. We are all persuasable, but not infinitely so. The most successful commercial activity seems normally to be that wherein a company identifies and/or creates a need which is perceived as worthwhile by the customer. Whilst it is always possible to argue that what is worthwhile can be an opinion induced by the manufacturer himself, there are probably many more examples where such an argument has failed than been substantiated, and there are certainly many better ways of making good returns on investment.

There is a more perplexing and complex situation which is often encountered in process industries, which Galbraith's preoccupation with the car industry perhaps led him to overlook. It is common to all industries which produce by-products in the course of manufacturing something perhaps totally in line with the accepted sequence, and for such organizations it would seem necessary, if possible, for the revised sequence to operate in respect of by-products. Alas, the evidence is that this is seldom the case. A classic illustration of the problem to which I refer is to be found at Arla Chemicals (1962).

As Figure 1 illustrates, the company's operations centre around the production of feran. All the other products the company sells were introduced in an attempt to make a profit from feran by-products. The by-products emerge in a fixed relationship to the volume output of feran. The problems within this complex industry are further aggravated by the ever-present danger of losing altogether the supply of raw material, baurite.

Until 1937, Arla had disposed of all its by-product sulphurous material simply by producing and selling sulphuric acid. The introduction of fertilizer products at this time was made largely because of adverse conditions in the sulphuric acid market. Since

Figure 1. *The Arla Chemicals Company*

Product and Process Relationships

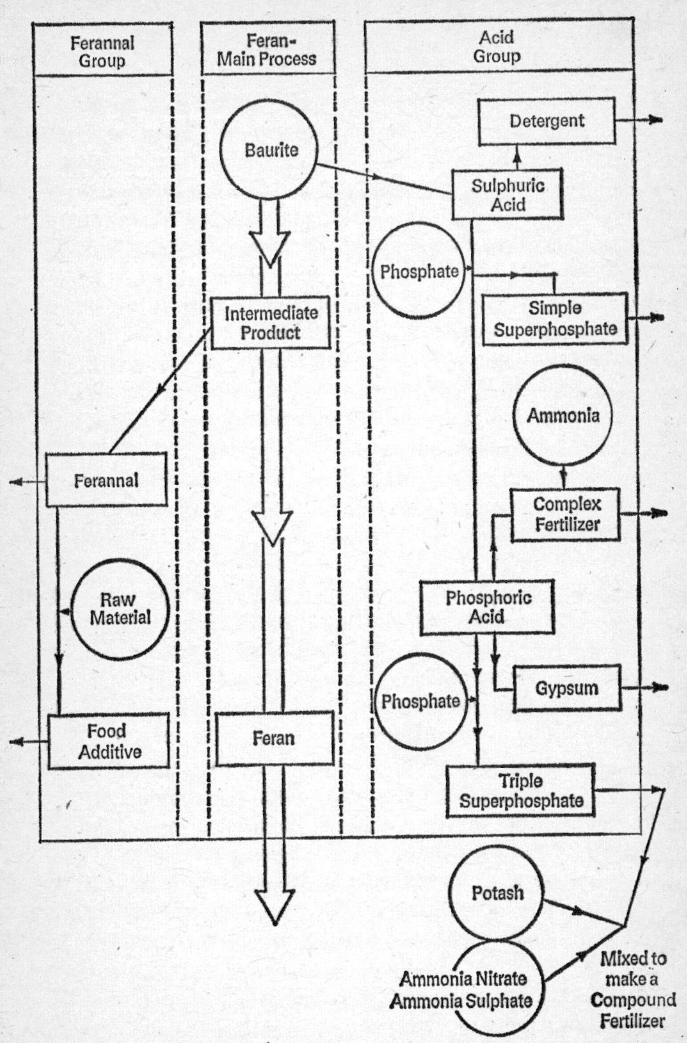

Source: Company Records

sulphuric acid had always been difficult to store or transport, other producers had had repeatedly to dump their output in the local market as supply temporarily exceeded demand. Consequently, the price for sulphuric acid had traditionally fluctuated greatly. Furthermore, as the world demand for feran increased substantially in the second half of the 1930s, Arla increased its production of feran and found itself faced with a greater supply of sulphurous material than ever before. At the same time, the general supply of sulphuric acid had outgrown the demand within the marketable area.

Under these circumstances, Arla began to process the sulphuric acid into simple superphosphate fertilizer, phosphoric acid and triple superphosphate fertilizer. These new products permitted Arla profitably to dispose of its sulphuric acid excess in new markets and also alleviated the problem of storage. Until 1962 Arla had been the only feran manufacturer in its geographical area that processed sulphuric acid into derivative products – in this case fertilizers.

After the Second World War the local pattern of industrial chemical requirements changed, and the demand for sulphuric acid greatly increased until it far exceeded the supply available from feran by-products in the area. Consequently, by 1962, Arla was producing one third of its sulphuric acid production from free sulphur to meet these needs.

Although the acid group consisted of ten products by 1960, only four of these were sold in any substantial amounts. Of these, sulphuric acid and triple-superphosphate fertilizer accounted for 80 per cent of the acid group sales. Simple superphosphate and complex fertilizers represented another 12 per cent, while the other six products accounted for the remaining 8 per cent. Phosphoric acid, one of these six products, actually was produced in great quantities, but it was primarily used as an internal intermediate product because no significant demand existed for it outside the company.

The markets for these products were diverse. The sulphuric acid was sold direct to industrial users within a 100/150 kilometre radius. The fertilizers were sold to farm supply distributors. Transportation costs limited this market to a radius of 150 kilometres. The other products were normally sold through agents in Europe.

The other major by-product, ferannal, was sold mainly to the food processing industry for use in the production of additives. However, Arla's customers had recently introduced a new process in which no ferannal additive was required.

I regret to say, and so I imagine do the executives of Arla Chemicals, that it is not at all clear how Arla can ensure that the philosophy inherent in the revised sequence would work itself out in such circumstances as these. But I concede I have taken an extreme example.

It is perhaps helpful if this examination of Galbraith's revised sequence is concluded by restating the manner of synergetic effect that is potentially available if technology and marketing can work sympathetically together. The cost of innovation can be reduced because the risk which separatism between marketing and technology necessitates that each must take can be reduced; the cost of innovation can be reduced because the time scale of innovation can be telescoped; the invoiced price to the customer can be reduced because the cost of innovation has been lessened. But none of this synergy can result unless both marketing and technology work together, each on the understanding of where its professionalism ends and where the professionalism of the other begins. The ideal company wishes to embrace the inventiveness of both professions and to judge not the source of the ideas but their customer-worthiness. There is more respect for the intellect of customers in industry than Galbraith gives credit for, although that respect has sometimes had to be learnt the hard way. Once again, Galbraith has omitted to develop the full implications of the customer backlash, but he did not fail to note it was inherent in the system he described.

THE EDUCATIONAL AND SCIENTIFIC ESTATE

Modern technology's basis in current knowledge and its desire for further development should enable it to avoid the worst consequence of Galbraith's technostructure, even assuming it to exist in its most sinister form as a conspiracy against customers. The educated and knowledgeable members of the very technostructure itself rebel against any social straitjacket inspired by the technostructure. They influence the determination of technological objectives, the pattern of organization to implement them, and the manner of demand management in specific sectors as shown in Figure 2. Furthermore, as discretionary incomes rise, they express their individuality by contracting out of the industrial system's mass market approach in favour of segmented modes of behaviour. Galbraith has identified this as no more than a possible method of protest – it could perhaps be more aptly observed as the most significant pattern of the future. As more and more join the educational and scientific estate, and more and more enjoy rising discretionary incomes, the creation and satisfaction of widely differentiated patterns of demand, as well as aesthetic and cultural objectives, can certainly be expected to become increasingly necessary. This is not a pattern of technostructure-managed innovation but of customer expectations. Of course, what can be conceived and created will remain the province of the creative mind, be it in research and development, or marketing, or customer's letters of complaint, or suggestion boxes.

Spurious differentiation of demand is and will continue to be cynically received by the educational and scientific estate. The most violent critics of detergent, cigarette and petrol advertising, for instance, can be found in the advertising companies and agencies themselves. Indeed, spuriousness has reached such a pitch that many customers are now treating a wide range of heavily branded products as no more than commodity goods – flour, petrol, detergent, beans, biscuits. This is one of the prime reasons for the massive burst (the last gasp perhaps of the spurious differentiators) in below-the-line promotion and own-label brands in the past ten years. Customers are not tempted into a new pattern of loyalties in such commodity markets. They merely take the marginal purchasing advantage proffered and move to any other national brand or own-label which next offers an advantage in its turn. As so many interviewed customers in these fields have remarked, a price cut would be preferable to them and it looks increasingly as if its effect would be no different from the manufacturer's viewpoint either.

One can surely not be surprised that the human mind should develop an immunity to bamboozlement. In face of it, we may identify our referent to fend off innovative indigestion. Stravinsky (1945) has neatly expressed these anxieties and the classic response:

> As far as I am concerned, I experience a kind of terror as I am about to go to work and, before the infinite possibilities offered to me, I feel that everything is permitted. If everything is permitted, best and worst, if nothing offers any resistance, every effort is inconceivable. I can't base myself on anything, and from then on every enterprise is in vain. . . . Nevertheless, I will not perish. I will conquer my terror and will take assurance from the notion that I have the seven notes of the scale in its chromatic intervals; its strong or weak beats are within my reach, and that I hold in this way solid and concrete elements which offer me as vast a field of experiment as this vague and vertiginous incident which has just frightened me. . . . What pulls me out of the anguish caused by unconditional freedom is that I always have the facility of concentrating on the concrete things which are in question here and now.

We could opt out altogether, using neo-luddite or hippie approaches to contemporary problems, or the nostalgic escapism of negritude as we go '. . . lumbering back to the clever tools [we] do not love and do not understand'. (Peters, 1964.)

Contracting out will not be necessary, however. Salvation lies within the industrial system itself even as described by Galbraith.

> In contrast with its economic antecedents, it is intellectually demanding. Men will not be entrapped by the belief that, apart from the production of goods and income by more progressively advanced technical methods, there is nothing important in life. . . . The industrial system brings into existence to serve its intellectual and scientific needs the community that, hopefully, will reject its monopoly of social purpose.

THE SOCIAL PLANNING OF TECHNOLOGY

Galbraith describes a contemporary market-place, where technology has replaced the customer as sovereign, technology not only in the shape of what can be achieved but also of what cannot. Although he hopefully sees salvation around the corner, he does not examine the powerful tools which society, as the representative

Figure 2. *Galbraith's Revised Sequence, with its linkage through the Educational and Scientific Estate*

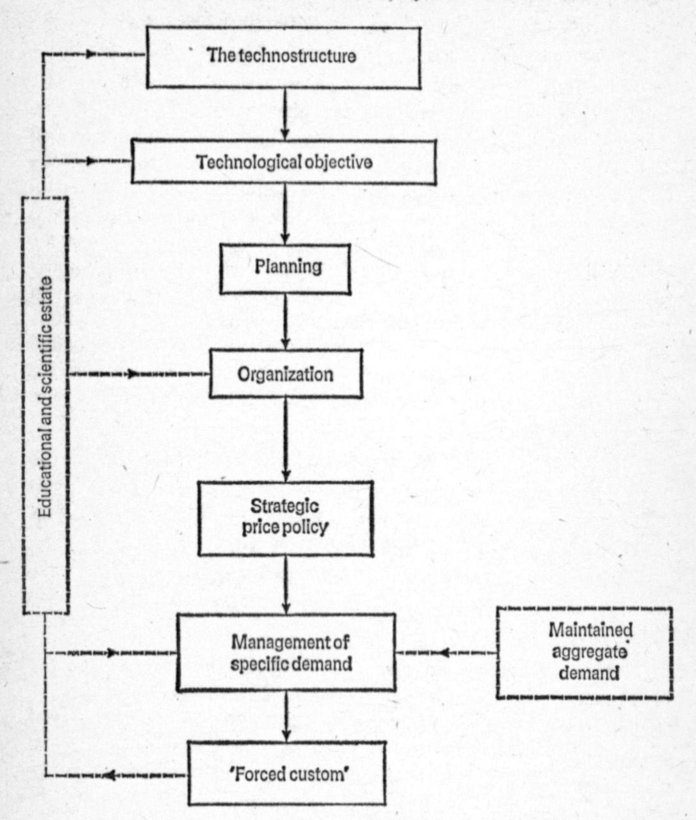

of customers, has at its command today. They go under the collective title of technological forecasting (TF) and although as yet by no means perfected, they show the way forward to a methodology for anticipating and hence shaping our technological environment. Managements have traditionally, albeit perhaps intuitively, kept a cautious eye on the pace of technological change in those sectors which they thought likely to impinge on their business interests.

But as is so frequently the case in management, it was a defensive eye more often than a purposeful one. The problem was often construed as the danger of being overtaken by new technologies before an adequate return had been obtained on a particular investment.

That this view is changed, and in many industrial sectors is now mere history, is reflected in a series of international, national and corporate events which have characterized industrial technology in the sixties. The issues have been pinpointed by the journalistic presentation of Servan-Schreiber, with his catchphrase, *le défi americain*; in the more sober thoughts of John Duckworth, managing director of our greatly expanded National Research Development Corporation; through countless political and ministerial pronouncements from the Wilson Labour government, beginning with Wilson's own 'white-heat of technology' speech, and its constant echo, first from Frank Cousins and, after his departure from Mintech, from Anthony Wedgwood Benn. The movement forward, *la sensibilisation* of senior management to a positive role for technology in our industrial prosperity, has been institutionalized not just in the early Ministry of Technology, but in our educational system, through the eleven new technological universities and the enhanced status of the polytechnics. It also became the major basis for our flanking strategy in the face of the Gaullist veto on British entry into the European Community.

This positive deployment of national resources as the basis for our industrial structure has entailed the abandonment of intuitive thoughts about technological futures. The new technology for the mastery of technology that was spawned, was not based solely on the extrapolation of existing trends in functional capabilities but on normative demands about the sort of future we wish to create; and as a vitally important corollary to that, on a realistic understanding of the sort of role we as a nation can play in the totality of world technology, either from the British industrial base or the wider industrial base of the European Economic Community.

The normative approaches inherent in much of the new field of technological forecasting are equally exhilarating from the individual, social and commercial viewpoints. They provide a rigorous framework for the exploration of alternative futures and the opportunity for all to participate more meaningfully in the development of the environment in which we and our heirs shall need to live. They provide the basis for the democratic management of the technostructure.

Within Britain already the purposeful structuring of our technological initiatives is beginning to pay off, although there are as many instances where the failure to take the initiative in technological change has led to catastrophic decay, both commercially and socially. The pattern of our trade with other nations, however, emphasizes some of the success we have achieved. Sir Fred Catherwood, a former director-general of the NEDO, has pointed to the transformation of the structure of our exports. From the situation in 1959 where 39 per cent of exports went to the sterling area, the comparable proportion in 1968 had fallen to 28 per cent. The shifting balance has been occasioned by the growth in our trade with the major advanced industrial economies in the world. We are moving towards a situation where we must increasingly 'live and cope with the rigorous competition of advanced high-technology, high-wage, capital-intensive economies, rather than the closed-trading system inherited from the empire'. Britain can no longer afford to cover the whole industrial waterfront. We must concentrate further in those areas of our greatest expertise. When this is done we can be seen to be succeeding, for example in aerospace, computers, electronics and chemicals. In each of these industries the techniques of technological forecasting are in action. The Programmes Analysis Unit, jointly established in the AEA and Mintech at Didcot, is also playing an important role in promoting their use both in strategic government planning and more widely throughout British industry.

TF gives us the ability to fight the future if we do not wish to become a victim of some particular facet which might otherwise be in store for us. Through TF society can lay bare the alternative futures and choose between them. The *Guardian* put the view succinctly, if in a somewhat pessimistic manner:

But as always, there is a ray of hope. The ideas germinated in the think-tanks may serve as an inoculation against the real thing. If we concentrate our thoughts now on 100,000-ton hovercraft, Mach-10 airliners, brains directly linked to computers, controlled longevity, and intelligent animals bred for low-grade labour, we may be able to resist them when they become practicable, as assuredly they will. It is bountiful of Nature to provide us with this defence mechanism. Caught unawares by concrete proposals to inflict such things upon us we might be unable to say no. And in that case the future of mankind would be nasty, brutish and everlasting.

TECHNOLOGICAL FORECASTING IN THE CORPORATE CONTEXT

The integration of the technological forecast into the framework of corporate long-range planning is fundamental to its effective use. It is directly analogous with the place of the sales forecast in the planning of operational activities for the company in the shorter term. Whilst the latter is now virtually universally accepted as the starting point for operational planning procedures, this is scarcely yet the case for technological forecasting in the long-range plan. This is primarily due to the ascendancy which the marketing function of the business has obtained during the past two decades. Whilst few would deny that the viewpoint of the customer had gone by default in the era of production and sales orientation which preceded marketing's ascent, the grip it simultaneously took on product strategy in many companies has led to increasing difficulties particularly in areas of rapidly advancing technology. Corporate planning demands the effective cooperation of the technologist and the marketeer if costly errors are to be avoided from both ends of the corporate process. The relegation of marketing to a realistic perspective is, however, a vital early stage in the development of effective corporate long-range planning. This relegation covers its usurpation of the strategic planning role in the company, not the process by which marketing's tentacles have reached out to collaborate in the process of new product introduction and control through a variety of organizational patterns such as task forces or new enterprise divisions. Without the process of strategic cooperation between the technologist and the marketeer, the company is doomed to rely on the often conflicting views of one or the other;

probably a recipe for less success in forecasting corporate futures now than in the past. There is an equally urgent requirement for the separation of operational marketing activity from its involvement within the strategic planning of the corporate future. These two strands are apparent in several emergent forms of organization structure in British industry, and are illustrated in Figure 3.

Figure 3. *Future/Present Dichotomous Structure*

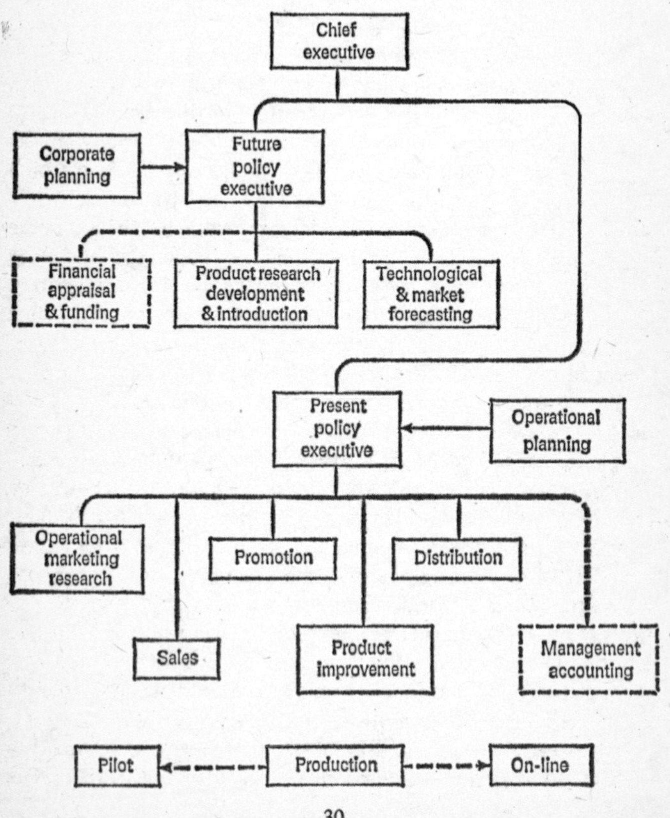

Product improvement and basic research are usefully separated, as are short- and long-term research into the market situation.

Nowhere is the need for technology/marketing cooperation more clearly demonstrated than in the recognition of the need for balance at any given time between the creating and satisfying of customer needs. No aggregated technological sector can exist unless its input technological sectors exist; equally none should exist unless its output satisfies a need either directly or as an input to another coexistent technological sector. Hetrick develops this concept fully into a model of technological planning, illustrating it with a technology composed of three sectors – aerospace, electronics and metallurgy (Figure 4). He assumes each exists at two levels (aerospace at pre-satellite and satellite levels; electronics as based on vacuum tubes and solid-state physics; and metallurgy as

Figure 4. *Technological Capability/Market Requirement Balance* (From Heterick)

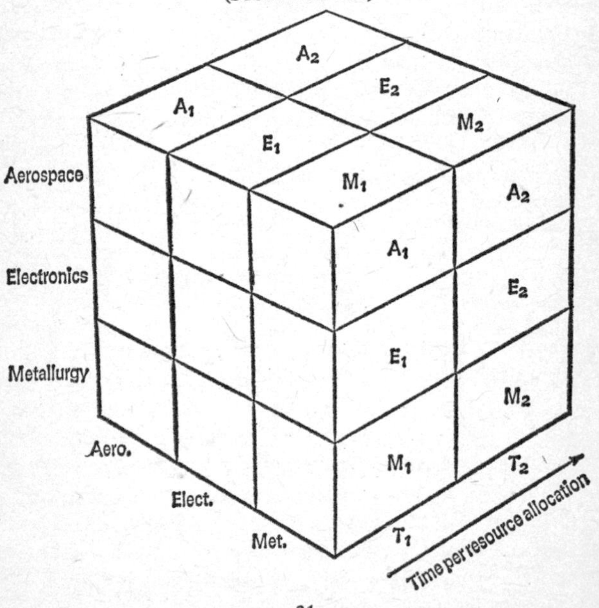

capable of producing 99·99 per cent pure and 99·999 per cent pure metals.) Figure 4 shows two balanced technologies to exist. Time is not, of course, used as an absolute here; rather it is relative to resource allocation.

Where new technologies are postulated for development, of course, such analysis indicates the sequence for development of component technologies. The analogy with networking is readily apparent. It has not been appropriate in these remarks on the corporate context of technological forecasting to comment on more than a few of the implications of technological change on company policy formulation. The full gamut is more exhaustively dealt with in Chapter 1.3. This key issue, however, has frequently been overlooked and has accordingly been emphasized at the outset.

One major problem remains as yet unmentioned. Positive optimism about a mastery of technology, based on our ability to forecast and plan, must return to the opening issues which were raised in the notion of technological myopia. Only a closely integrated relationship between technology and marketing can ensure effective technology transfer and a profitable balance between market requirement and output technology. Gruber and Marquis have focused attention on the variety of elements making for successful transfer, and suggested as a model:

Magnitude of transfer = f (source, nature of item to be transferred, structure of channels for transfer, potential recipients, i.e. customers, of items to be transferred)

Each element of this transfer structure needs careful examination, both separately and as an element in the total transfer process.

Both these issues – of technology transfer and the balance of technologies (especially aggregated technologies) with market considerations, go beyond the normal scheme of TF. Allied to TF, however, they provide a powerful way forward to an intelligent restatement of the marketing concept in the overall corporate context, which avoids the technological myopia which has at times looked like becoming hereditary in the marketing family.

The interaction between the two hitherto largely discrete areas of the business, marketing and technology, will characterize the form of presentation adopted in this book. For the research and development scientist such a stance may be unfamiliar or indeed uncomfortable, but it necessarily is the pattern for much of our business future in this century.

PART ONE

THE MANAGERIAL IMPLICATIONS OF TECHNOLOGICAL FORECASTING

CHAPTER 1.1

The Technological Setting

DECISION-MAKERS have always wished to know what the future holds in store for them, and what the effects of their decisions will be in order that they might evaluate the alternative courses of action open to them. Within every business decision, some estimate of the future is implicitly made, and the question arises whether such an estimate should be made unconsciously as an implicit part of a decision, or whether it should be deliberately arrived at and specifically stated. In the latter case the validity of the estimate can be tested, since the method, data and premises used in arriving at a forecast are revealed.

With an increasing rate of technological change, firms have come to realize that the expensive and far-reaching changes they are called upon to make involve many interrelated factors. To evaluate each, along with its impact, requires consideration and coordination of a number of economic, technical and social elements, together with their projections into the future. Planning the future involves predicting the decisions that policy-makers must take. This inevitably involves a high degree of risk, especially when carried out on an informal basis, as decisions depend upon quirk and temperament as well as reason. Even if reason alone were at play the task would not be that much simpler.

From a technological point of view, issues of the future that are raised include: accelerated nuclear proliferation; loss of privacy, greater government and/or private power over individuals; over-centralization; decisions that are too complex and important to be left to single individuals; new capabilities that are so inherently dangerous that they may be disastrously abused; too rapid or cataclysmic a process of change for smooth adjustment, and so forth. Traditional products, skills, materials and production facilities may become obsolete within a short span of years, and in some cases within just a few months, with the result that many thousands of businesses will rise or fall on their ability to respond effectively to an accelerating technological challenge, affording both threats and opportunities.

The first requirement for any businessman under such circumstances is a keen sensitivity, awareness and receptivity to technological change as a major environmental force which he can employ, and to which he *must* respond. In many firms it will be by far the most potent force, and consequently technological progress must somehow be anticipated.

Technological forecasting (TF) has emerged to fill this role because it was urgently required. Although the concept of formal technological forecasting is a complex one, it is vital to future corporate success. This can clearly be seen from the help it accords planners in identifying features of the future environment that are of potential interest to their firm; in identifying research and development (R & D) projects as future opportunities; in influencing and guiding shorter-range plans; and in allocating resources amongst the various functions of the firm to anticipate, counter, or assimilate competitive technological innovations.

The successful firm will need not only to perceive opportunities, but also to anticipate them. But the problem following from anticipation is that of acting correctly until that anticipation becomes reality. It is in this respect that technological forecasting aids management in developing the necessary strategies and deriving from them suitable technical objectives in order to optimize the deployment of resources and techniques.

Technological progress and innovation must nowadays be regarded as something quite normal rather than unnatural. Future success is hardly likely to come to the firm that believes technological change to be the enemy of orderly, planned activity. However, rapid change and technological innovation have profound effects on human society as well as on business organizations. As a result, the value of such change should be measured in terms of increased efficiency, be this in time (for example with pressure cookers), size (transistors), or motion (motorized transportation). The major significance of technology from the human standpoint is in its ability to make human effort more effective and harmonious.

Accordingly, the firm must be able to assess the impact on

society of technological progress. Progress that is incongruent with human values, as a result of a lack of technological foresight, is the outcome of a false sense of security about the effects of technological change.

It is important to appreciate that technological change is but one aspect of development, and that development is normally a historical process with different stages each having its own partial but definite goals. True innovation embraces much more than technology. From the corporate viewpoint, it includes changes in production, finance, administration, marketing, training and much else besides. To seek technological innovation or development without considering improvements in the other elements of the enterprise leads the firm to an unbalanced position. In this sector, TF can help to achieve integration across formal organization lines.

The basic problem in TF is making decisions *today* with only uncertain knowledge about the future, both as regards the facts and the human value systems. For long-term planning purposes, the identification of possible future innovations is insufficient in itself. The future social, economic and technical environment must be forecast in a coherent way. Since our own coherent appraisal of the present is still rather poor, such an appraisal of the future is bound to be fraught with difficulties. None the less, only by forecasting expected future conditions can progress in the multifarious dimensions of modern life be effectively planned. The alternative is for inventors to run the risk of creating what society is unable to use, understand or afford at the time it becomes available.

A good deal of imagination is necessary on the part of the forecaster, and this is increasingly true as the necessary time scale for forecasts increases. To trace out distant objectives along a direction which remains the same is an exercise which gives the illusion of foresight, while in fact being almost meaningless since it simply suggests that we shall be going on doing much the same sort of thing. The problem is serious, since, if we do not proceed in accord with current patterns, what are the realistic alternatives? Our answer to this question lies very much within the forecasting techniques adopted. We must seek to temper extrapolative interpretations of the future with normative forecasts, and we shall explore the methodology fully in Part Two.

Long views must be taken, and the longer they are, and the more significant their consequences, the more vital it is that they should be rational. Any forecasts underlying decisions about the future must be examined in the light of the probability of their being borne out. Such an examination will rely on both intuitive and systematic forecasting methodology. To decide the correct combination of these two is a problem facing every decision-maker.

The alternatives from which management must choose for future development are largely technology-determined, and the choice between several types of logical futures and decisions is often subordinated to technological ends. Such factors may be systematically predicted, but the notion of desirability must first be conceived purely in human terms, relative to a future state, and only after such an independent image has been created can considerations referring to technological attainability be introduced. In this way, *the important point becomes that of changing the present to fit the image of the willed future, rather than protecting the present along the logical vectors that happen to be inherent in it.* Without doubt it is a wrong allocation of energy that results in an obsession with feasibility and a neglect of desirability. Such a posture is tantamount to a paralysis of our ability to ascribe correct values to our goals.

THE EFFICACY OF TF AND ITS LIMITATIONS

Technology has today become perhaps the most important factor in social change and substitution. If allowed to develop in an uncontrolled way, it frequently leads to unstable solutions to problems, to disequilibrium. Instability in the dynamic system of society follows with accompanying waste and uncontrolled or cancerous change. Within the narrower corporate context, the outcome of unregulated technological development often means that:

(i) technological feasibilities, when recognized, are turned into reality: *can* is understood as *ought*;

(ii) market demand is followed blindly;
(iii) product lines are pushed to their extreme; and
(iv) demand is created artificially for the sake of specific scientific and technological developments, or to maintain skills and capabilities.

In general, uncontrolled development will occur when a deterministic attitude is adopted. This can be checked by the use of TF. To be useful, such forecasts do not necessarily have to predict the precise form technology will take at some specific future date. As with other forecasts, their purpose is simply to help evaluate the probability and significance of various possible future developments in order that management may take better informed decisions. The efficacy of TF is most clearly seen in its substitution of probability for determinism.

Forecasts, however, are useless, no matter how sophisticated, unless they eventually influence action. Furthermore, even if they do so influence action, their efficiency is critically dependent upon the goals that are sought. The range of TF techniques that has been developed and tested in industrial settings, enables management to analyse such issues as:

(i) the demand forces calling forth a particular technology;
(ii) the ultimate theoretical potentials and relevance of that technology;
(iii) its expected rate of progress; and
(iv) its probable impact on the company and the society.

Technological development takes place in a dynamic environment that strongly influences the course of that development. TF therefore begins with an evaluation of foreseeable economic and social changes, and how they are likely to modify the level and direction of technological development. It is the responsibility of the forecaster to ensure that the impact of TF on social and economic forecasts leads to internal consistency of the total result.

One useful method that allows a distant projection into the future is that of inventing alternative futures, termed scenario writing. This consists of constructing as many internally consistent, plausible and significantly different scenarios of the course of future events as possible and then deducing from each the probable consequent course of devélopment in relevant technological areas. Other methods in TF include extrapolation and morphological analysis. Each of these are described in detail in Part Two. Once technological opportunities and their allied time-scaling have been indicated by any of these techniques, the managerial input for the selection of actionable technologies has been provided. When such directions for action have been set, they must be kept under constant review in the light of changes in technology or competitive initiatives.

TF techniques involve a high degree of complexity, owing to the need to forecast not only the discoveries of science, and when they will happen, but also to enable estimates to be made of their repercussions in a wider framework. This means that TF must be regarded as a continuous process.

Complications really start, however, when the forecaster attempts to pull the strands together. Mathematics is helpful, especially with computer application, to derive important relationships. However, mathematics cannot foresee discontinuities, and may even fail to predict a continuous process if some relevant facts have been omitted.

Help can come from various logical devices, such as three-dimensional diagrams (with, for example, economic development, time, and technical ability as the three axes) and tree and matrix analysis. Such devices can serve an important role as thought orientators. A three-dimensional diagram may suggest that it is economic growth rather than technical refinement that dominates a product's market potential, and this will aid in controlling thought in the appropriate direction.

A knowledge of the interrelationships between science and economics, and within science itself, shows the importance of not attempting to proceed with a new product until all the necessary techniques and market conditions are on the point of realization. Jet engines, as an example, could not be developed until metallurgy had come up with alloys able to function in hot jet gases. The world is learning the hard way that building a supersonic airliner is more than a matter of knowing how to fly fast. It requires the contribution of many technologies. The idea that progress can only be as fast as the slowest relevant component technology is an unwelcome one, but it is a useful guide to where future research effort should be directed.

TF is an exercise in logical deduction as opposed to straight mathematical calculation, and demands a substantial input of imagination, as does science itself. Its biggest incentives are commercial in nature, but as industrial concerns become larger and more capital-intensive, the cost of mistakes rises. The choice between risk-incurring innovation and the apparently low-risk approach of doing nothing is one of the great dilemmas facing modern management. Innovation, so often lauded as being in the public interest, must be exploited through adaptive management. One cannot regard change as predetermined, and hence foreseeable.

There are considerable difficulties involved in convincing managers unaccustomed to using TF techniques that they are of value, even essential, in making critical decisions. Guidelines for the adoption of TF include working towards a realization that forecasts should develop the pragmatic insights needed to make this year's decisions. They need not focus on esoteric problems of the year 2000. Distant events often fail to carry managers because they lack reality unless they actually impinge on current decisions. TF forecasts must place opportunities and threats in an appropriate order of priority, and integrate them within the firm's regular cycle of executive decisions. It is desirable, wherever possible, to expose promising executives to technological planning and forecasting activities as a routine part of their training.

The adoption of an interdisciplinary approach, i.e., the widening of the data base by the inclusion of social, political, biological and psychological parameters, will be insufficient by itself to ensure the right use of TF. It may merely ensure a more sophisticated style of societal manipulation towards goals which have been chosen by a power élite. A distinction must continually be drawn between organizations with a common purpose that must be achieved by the members together, and those whose members are engaged in separate pursuits without collaborating to attain a common objective.

No matter which techniques are adopted, TF must not be perceived of as a development panacea. There are some important limitations. The first relates to unpredictable interactions. The interaction of several discrete technological advances may create totally unexpected potentialities which can shatter all forecasts.

Secondly, unprecedented demands may arise. Completely unforeseeable future conditions and events may occasionally create whole new areas of primary and secondary demand. Dramatic new technologies will continue to have self-amplifying effects on demand; but an imaginative conception of product use and formal planning to supplement the product's initial demand cycle can help to foresee some effects of this sort.

Major discoveries provide the third limitation to TF. The discovery of totally new phenomena may open the way to significantly new potentials in the way that transistor and laser technologies have.

Finally, inadequate data constitutes a severe limitation to many managerial functions, not least to TF. It is currently the single most limiting factor in the development of better TF. This inadequacy of source data may make it necessary for the forecaster to develop his own primary data before proceeding to analysis. Cost considerations will generally limit the relevant populations the analyst can sample, and the accuracy of his studies can thus be affected.

The rate of technological change has been accelerating on an international level, especially during the last twenty years. This has meant a transformation of the fundamental character of technology, with new technologies, e.g., nuclear power, computers and satellite communications, added to the older areas such as electrical and mechanical engineering, textiles and chemicals. The pace of technology-transfer is also increasing with the greater development of inter-disciplinary skills.

The proliferation of innovations and ideas results in a fluid and exciting business environment, with the unexpected an ever present element in the game, and flexibility the fundamental rule. An air-supported bearing, a solid-state device, welded fabrication, powder metallurgy – each can change the whole perspective of an industry, suddenly or with little warning.

Larger firms in Europe and in North America already have TF teams at work. The Xerox Corporation has set itself a target of $2,000,000,000 turnover by 1975. In order to obtain half this future figure, the corporation is relying on the generation of new ideas by TF. The British government's Programmes Analysis Unit at Didcot is forecasting developments in water supplies and submarine communications. Current initiatives are carrying the techniques of TF into broader fields, so that social problems such as population drift, educational deficiencies and pollution can be anticipated and avoided.

Continuous and rapid development has certainly been a feature of the situation in computers. Over the past decade the speed of the fastest computers has increased from 2,000 operations per second to 500,000. For the typical manufacturer, the cost of 1,000,000 additions has fallen from £5 to 12p. So far as anyone can see, this rate of development will continue and indeed accelerate – a fact which has dramatic implications for white-collar/clerical staff.

It will become economical to harness computers for more and more purposes, but as hardware costs fall they will be swiftly offset by rising software costs. We can readily forecast widespread development of programming in plain English or algebra – in fact, more programming by actual users such as accountants or sales managers, and less by professional programmers.

The essential contribution that computer technology can make to management is by way of improvements in the communication and handling of information. It is possible now to foresee on-line management systems in which production, sales, distribution, accounting and the rest are all linked by the exchange of information in the form of streams of electrical impulses flowing between computers. Techniques are being developed so that engineers or architects can draw their designs directly on to television-like screens of a cathode-ray tube, and have all the dimensions stored automatically by the computer for recall at some future time. A start has also been made on coding and storing the law, thereby aiding in the search for precedents, and in revealing any overlaps or contradictions. Thus professional as well as clerical employees will be significantly influenced by developments in computer technology and usage.

It has been confidently forecast that by the year 2000 science and technology may have provided: weather control over whole regions, with deep implications for the recreation and clothing industries; chemicals to stimulate the growth of new limbs and organs; and aircraft able to travel from London to Sydney in one hour. Direct communication between the human brain and the computer may also be achieved; intelligent animals may replace low-grade human labour; and unlimited power may come from the reactors which harness the H-bomb.

In applying their art, for art it is, technological forecasters will not try to predict what the world *will* be like, but what it *could* be like. They will, accordingly, try to work out technologically feasible options to serve as a basis for choice, thus enabling man to shape, rather than suffer, his destiny.

Gabriel Bouladon, of the Institut Battelle in Geneva, has predicted that, by the year 2000, all motor vehicles will be banned from city centres, and will be replaced by conveyors moving at 7–10 m.p.h. Outside the cities, electric cars will become important, and the impact on the motor industry will be severe. The impact of such possible social decisions must be fully explored. He also anticipates that the main competition for road transport could be from vehicles totally enclosed within a tightly fitting tube, and floating on a thin film of air. They could be driven at speeds up to 500 m.p.h. by intracting the air in front of them, and pumping it out behind. This could be the most economic form of transport ever invented, with serious consequences for certain sectors of the petroleum industry. Cargo ships could be powered by nuclear energy, and air traffic may well increase by 8,000 per cent in the next twenty years. And our foodstuffs in 2000 A.D. will probably be protein produced by bacteria and other micro-organisms living in organic wastes and petroleum fuels, unless we decide to prevent it.

In concluding these observations on the overall setting of TF, it is worthwhile to focus on the proposition that all decisions are decisions about what *should* happen in the future, and are based at least implicitly on forecasts of what *will* happen in the absence of any decision, and in the event of each variety of alternative sequences emerging. The more thought that can be given to the forecasts in the time available before any final, irrevocable decisions have to be taken, the better. But above all, no matter how much thought is given to forecasts, once made they must continually be treated with scepticism.

CHAPTER 1.2

The Environmental Impact

TECHNOLOGICAL change acts upon and interacts with a wide range of environmental phenomena. In the industrial context this environment can be modelled as in Figure 1.

Figure 1. *Environmental Factors Encompassing the Firm*

Whilst Figure 1 does not present an exhaustive inventory of environmental factors, it is intended to make apparent the myopia of anyone who focuses solely on technology. It must be treated as a tool of society and the individual corporate enterprise; the two cannot realistically be considered apart. Indeed, if the technological forecaster ignores society, he in turn runs the risk that society will ignore his forecasts. Social pressures may even override clear-cut technological recommendations, such as bacteriological warfare, in favour of less potent but socially more acceptable alternatives. Technology might not always do well by society, but whenever a technical invention crosses the societal threshold, it can only do so at the bidding of a human manipulator.

The emergence of sustained, large-scale and time-compressed instances of human manipulation of technology has been a comparatively recent phenomenon. It was not, for instance, a feature of the first industrial revolution in the eighteenth century, although the magnitude of the societal impact was perhaps even greater then.

Within this setting, the technological forecaster must fully appreciate that the social system has multiple goals which can only be attained through cooperative effort. Perhaps the most obvious example of a societal goal is national security; but most people today will include health and education. Survival is, indeed, possibly the highest goal of mankind. Adaptation to the social and non-social environment is obviously necessary if goals are to be attained. This process of adaptation can be seen to be two-way as far as social systems are concerned; any change agent must normally adapt the environment to its goals, and also adapt its goals to the environment. The ultimate goals within such a process are our social values, a good Western example being political freedom within a mixed economy.

In addition to this adaptation, human and non-human resources must be mobilized in an effective way, according to the specific nature of tasks. There must, by way of an illustration, be a process for ensuring that enough persons, but not too many, occupy each necessary role at a particular time; and also a process for determining which persons will occupy which roles. Such allocation of members of the society must then be supplemented by the allocation of scarce, hence valuable, resources if both goal attainment and adaptation are to come about. The difference between the two is a relative one.

Full-employment may be deemed as a high priority societal goal, but this in its turn is greatly influenced by technology. Since 1900 many new products have appeared, such as television sets, aeroplanes, motor vehicles, vacuum cleaners, and refrigerators. If there were no television sets, one might wonder whether or not all those currently engaged in their manufacture would now be unemployed, or be gainfully employed elsewhere. Technological change undoubtedly creates new employment opportunities; but it also results in obsolescence of older products and techniques, and increasing automation involving the replacement of men by machines in certain types of work. Both aspects have a strong bearing on the social objective of full-employment, and a wider effect can be seen in terms of inflationary trends in this context.

Sheer *success* is also frequently seen as a desirable social goal, and various incentives for success are provided by established cultural values. However, extreme cultural emphasis on the goal of success may and often does result in the use of any means to achieve this goal. This is the process of 'demoralization' in which norms are robbed of their power to regulate behaviour, and anomie (the breakdown in the cultural structure) ensues. But this is an extreme occurrence, and in the normal course of events success will be found to be reinforcing. After a certain amount of success, goals which formerly appeared impossible may suddenly appear possible. In this way success can reinforce not only higher specific goals, but also a stronger concept of one's own power to reach them.

It is fruitful here to compare the relative patterns of success-striving in North American society with those in Western Europe. The relative incidence of private affluence and public squalor in North America and Western Europe seems to bear witness to the sacrifice of societal goals by the former in its race to achieve outstanding technological advancement. Nowhere is this more apparent than in urban development, especially the lack of town and countryside planning and the pollution of the landscape of the United States and Canada. But the ending of the sixties has seen an awakening interest in this which looks for much of its leadership to the cultural and social heritage of Western Europe in precisely the same manner as Western Europe looks to North America for much of its technological advancement. Whilst men have reached the moon from United States soil, there is greater poverty and maldistribution of wealth and welfare in that land than in other less affluent societies.

Galbraith has argued from his North American background that what really counts is not the quantity of our goods, but rather the quality of our lives. This we must relate to human needs. Much of the world's and of our national population still has unsatisfied basic needs. If this situation is to be remedied, and the standard of living and way of life of the community at large improved, the business world must play a leading part. The projection of existing consumption patterns does not, however, give any indication of the likely effect of new products, new promotions, and new processes on the patterns of future consumption.

The problem of finding out what will be needed in the future is more than just a problem of adding up the figures for what is lacking today; it involves the exciting and demanding task of conceiving what it is that would be wanted if only it were invented or could be imagined.

Thus, working from a given goal, either socially or technologically determined, the technological forecaster must review undeveloped as well as existing forms of technology.

The role of the technostructure, in Galbraith's analysis, influences much of our life, and nearly all of it that involves earning money and spending it. The technostructure sets our prices, influences the pattern of our purchases, and distributes the resulting income to those who participate in production. Its planning extends within limits to the management of the demand for those products that are purchased. The outcome is that to know now and to what ends our futures are governed, it is necessary to understand the goals of the technostructure and of the individuals who make it up.

The problem of goals in this context begins with the relation of the individual to the organization, in this case the technostructure. That which the organization seeks from society will be a reflection of what members seek from the organization. Similarly, what society can ask from the organization will, in turn, depend on the relation of the organization to the individual.

It has been observed that humans, in contrast to machines, evaluate their own positions in relation to the values of others, and come to accept others' goals as their own. But an individual, on becoming associated with an organization, will be more likely to adopt its goals in place of his own, if he has hope of changing the goals he finds unsatisfactory. It follows that if a group, such as the technostructure, holds power, it will be able to adapt, within reason, the goals of the enterprise to its own. This is a strong motivating force, and it is imperative that in acting in this way the technostructure identifies itself with the goals of the nation or the community. If this latter condition is not met, a lack of consistency between the goals of society, the organization, and the individual will emerge accompanied by a similar lack of consistency in motivation. In this matter, which is a social one, we have a deeply interconnected matrix, following from which, if the goals of a society are known, guidance is given to the goals of organizations serving it, and also of the individuals comprising them. The reverse also holds true.

Galbraith's disturbing sequence of events and ideas includes the suggestion that the goals of a mature corporation will be a reflection of the goals of the members of its technostructure, and that the goals of the society will tend to follow upon those of the corporation. If a society values technological virtuosity highly, and measures its success by its capacity for rapid technological advance, this will indeed become a goal of the corporation, and thus of those who comprise it.

Identification is an important concept in this respect since the individual will only identify with the corporation if its goals, as he sees them, accord with some significant social goal. In this way the firm making life-saving drugs will win loyalty and effort from the social purpose its products are presumed to serve.

In the American economy, those engaged in the design or manufacture of space vehicles will identify with the goals of their organization because they are in accordance with the scientific task of exploring space, or the high political purpose of outdistancing the

Russians. Similarly, the manufacturer of an exotic missile fuel, or a better trigger for a nuclear warhead, attracts the loyalty of employees by being seen to serve the cause of freedom. There can be no such wholehearted identification if the firm is simply engaged in making money for an entrepreneur, and has no other claimed social purpose.

This process is deceptively circular in as much as what is believed to be socially important is often the adaptation of social attitudes to the goal system of the technostructure. This latter goal system principally revolves around the production of goods and associated activities. It is important to them that this function be accorded high social purpose. From a detached viewpoint, however, the continued or expanded output of many goods is not easily accorded this social purpose. More cigarettes cause more cancer; more alcohol causes more cirrhosis; more cars cause more accidents, plus more pre-emption of space for roads and parking, along with more pollution of the air and countryside. From the opposite standpoint, increased production leads to growth, and the growth of the firm as a goal of the technostructure is strongly supported as a social goal because it aids overall economic growth. The concommitant annual increase in gross national product (GNP) will be unanimously applauded. It provides the potential wealth to alleviate suffering and hardship and to help progress towards priority social goals. The part technology plays in economic growth requires little elaboration here. Most of us today clearly value technological advance itself, and see it as a provider of much social good.

If Galbraith's thesis holds, it would appear that the technostructure is no longer primarily responsive to pecuniary motivation, but rather sees itself as identified with social goals, or with organizations serving social purposes. However, aesthetic goals are beyond the reach of the technostructure in the sense that it cannot identify with them. Hence, if economic goals are thought of as being the only goals of society, then it is natural that the industrial system should dominate the state and society, both of which should serve its ends. Yet when other goals are strongly asserted, the industrial system will fall into place as a detached and autonomous arm of the state, responsive to the larger purposes of society. The implication of major importance for the firm acting to catalyse technological change, and engaged in TF, is the imperative need to integrate its view of the several goals of the state, the social system, and the industrial system in 'designing the future'. If social disaster is to be avoided, balance must exist.

Once the goal can be determined, the action problem is purely cognitive, with the solutions being subject to the primacy of cognitive standards, with appreciative and/or moral considerations being subordinated. When the goal is not specifically given, appreciative and moral considerations must be allowed to play their part in goal identification. Difficulties will arise over certain species of goals – especially those of a non-empirical nature, such as aesthetic goals. One of the few instances in which an organization has been dominant in the realization of a non-empirical goal is the Roman Catholic church.

Four types of deviation from goal-attaining norms can be isolated:

(i) striving for prescribed goals using proscribed means;
(ii) using institutionalized means but failing to strive for prescribed goals;
(iii) rejection of both institutionalized goals and means; and
(iv) deviation from cultural prescriptions defining both means and goals by excessively conforming to each.

Such possible deviation must be consciously checked in process, lest the firm's activities and development end up only tangentially related to general development.

INTEGRATION OF SOCIAL AND TECHNOLOGICAL CHANGE

Technology-related changes have penetrated virtually all areas of social life in such a way that mature individuals will now have experienced several significant technology transitions within their own life-span.

The fact that these changes have occurred relatively precipitously, within the easy memory of individuals, at a rate that does not permit gradual, unconscious adaptation to them, forces upon us an awareness of change that undermines our faith in the stability of any future socio-economic state.

As a result, great care must be taken to ensure that social and technological change are fully integrated. Yet there is seldom any really clear perception of this integrating process. If we could freeze technology, society would continue to evolve subject to that restriction. And if we could freeze society, technology would continue to evolve subject to that cognate limitation. One may reasonably ask, therefore, that if each is free to evolve in a world in which the other is fixed, what links them together? Unfortunately, there is no single or definitive answer.

Oto Sulc, during a six-month period as senior research fellow at Manchester Business School in 1968, conducted some pioneering work in this area. He started with two premises. First, that the optimal exploitation of human and national resources depends upon knowledge about the extent to which progress in technology is likely to be affected by changes in the future social environment, and vice versa. Second, that such an understanding is basic to the choice of whether, and in which direction, to adjust investment in technology to the existing social structure, or to attempt to modify the social environment better to accommodate changes in technology. Sulc aimed to develop a methodology for the description, in both quantitative and qualitative terms, of the interaction between anticipated technological and social change.

He explicitly acknowledges that the treatment of these matters requires considerable cooperation between forecasters in the fields of both physical and social technology. Whilst a good deal of attention has been paid to methods for the forecasting of technological change, very little has been given to methods for forecasting social change, and even less perhaps to the interaction of the two.

Figure 2 presents managerial opinion on the relevance of measures intended to mitigate the detrimental effects of various developments predicted to occur in computer-controlled technology. The list of measures is designed to influence the behaviour of employees who will be affected by the changes in the computer technology.

Figure 2. *Relevance Scheme of Interrelationships between changes in the adaptability of management and effects of development in computer control (from Oto Sulc)*

Effects of future trends in computer control engineering the resistance of management → / Measures improving the management adaptability to future trends in computer control ↓	Relevance in Process Technologies					Average relevance in all industry
	1. Demands for qualification and personal dispositions	2. Redundancy of middle managers	3. Loss of responsibility and decision autonomy	4. Impersonality of procedures and communication	Total relevance	
1. Training of operational management staff in computer application, throughout all aspects of the firm's business likely to be affected by computer technology starting from junior graduate posts	2	3	1	1	7	high
2. Acquiring operational management staff so trained in computer application from outside the company, particularly from Research Institutes	2	0	0	1	3	minimal
3. Development and application of methods for selecting computer system managers and scientists with appropriate intelligence and personal characteristics	3	0	1	0	4	minimal/ moderate
4. Inter-company agreements on the transfer of displaced middle managers	0	3	0	0	3	minimal
5. Inter-departmental transfer of displaced middle managers to newly created jobs within a company	1	3	0	0	4	minimum/ moderate
6. Inter-departmental training and experience for middle managers before introduction of computer.	3	0	2	2	7	high
7. Changing the social prestige image from membership of the decision hierarchy to that of a technically expert élite	2	0	1	0	3	minimal

Figure 2. *continued.*

Effects of future trends in computer control engineering the resistance of management → / Measures improving the management adapatability to future trends in computer control ↓	Relevance in Process Technologies: 1. Demands for qualification and personal dispositions	2. Redundancy of middle managers	3. Loss of responsibility and decision autonomy	4. Impersonality of procedures and communication	Total relevance	Average relevance in all industry
8. Greater adaptability of computer programmes to conform to the varying qualifications of operators and programme users (for example, by more flexible sub-programmes, and more natural computer 'language')	3	0	2	1	6	moderate/ high
9. Designing the technology of computer maintenance and service with the aim of employing a greater number of semi-skilled ones	3	1	0	3	7	high
10. Changing the social prestige image of managers in the environment as a whole, from decision and control functions to coordinating functions.	0	0	2	1	3	minimal
11. Facilitating and simplifying flow of information between: a) highly specialized groups or departments (horizontal flow of information)	1	0	0	3	4	minimal/ moderate
b) between staff and line functions such as systems design, computer personnel and operational management (for example, by the introduction of suitably qualified liaison officers)	7	0	0	3	5	moderate
TOTAL RELEVANCE	27	10	9	15	56	

Figure 3. *Average efficiency of social changes in improving workers' adaptability to future technology (from Oto Sulc)*

Effects of future technology → / Measures supporting workers' adaptability to technological changes ↓	1. Increasing tempo in changes in qualifications; 2. Insufficient interchangeability of workers; 3. Insufficient mobility of labour; 4. Displacement of lower paid jobs
1. Gradual retraining of workers	high
2. Intra-company agreements on the transfer of technical personnel	moderate/high
3. Legislation lowering the retirement age	moderate
4. Unemployment insurance approaching the national minimum wage level	minimal/moderate
5. Public retraining centres	minimal/moderate
6. Government supported companies' retraining centres	minimal/moderate
7. Development of greater use of aptitude and IQ tests for selection of operators	minimal/moderate
8. Transfer of the collective bargaining to the shop-floor level	minimal/moderate
9. Leaving old titles and old job evaluation	minimal
10. Radical retraining of workers over 55	minimal
11. Raising the school leaving age to 18	minimal
12. Creation of easy types of jobs for older workers, etc.	divergence of opinion
13. Development of favourable attitudes to the status of such people	divergence of opinion

The measures supporting the workers' adaptability are related to the effects of future technology. The average effectiveness of proposed measures was estimated on the basis of opinions of managerial experts, which were gathered through questionnaires. The potential effectiveness has been related to a common denominator of all four technological impacts.

Sulc concludes, on the basis of Figure 3, that a firm operating in the consumer-product field must take note of the market and technical demands irrespective of whether its workers are adaptable or not. If they are adaptable, the move will go smoothly; if they are not adaptable they may well be faced with extensive redundancy. Moreover, low adaptability on the part of workers to technological changes may be hypothesized as a frequent explanation of lower-than-designed efficiency in systems. Some estimates of non-utilization of manpower in the UK suggest that the figure is as high as 20 per cent.

Finally, workers' fears of being unable to adapt to technological changes result in trade union or professional resistance to quite plausibly effective innovations in British industry. Automation in the newspaper industry and containerization in the docks and railways are recent classic examples.

Successfully integrated planning in the areas of expensive and sophisticated technology often requires the state to underwrite many costs, including some of the costs of research and development, and to ensure a market for the resulting products. It is important, therefore, to the technostructure that technological change of whatever kind be accorded a high social value. In consequence, the underwriting of sophisticated technology by the state *has* become an approved social function. Few people question the merit of state intervention for such social purposes as supersonic travel or improved applications for nuclear power. Even fewer protest when these are for military purposes.

The Labour government of 1964–70 continually extended the role of its Ministry of Technology, particularly in October 1968. Mintech was eventually established with a central command over all industrial policies – for new and developing industries, for investment policies in both the public and private sectors, for commitments by the state in private firms, and for research and development through the Science Research Council, National Research Development Corporation (NRDC), the now defunct Industrial Reorganization Corporation (IRC), and several other agencies.

In terms of goal attainment, the government, as society's agents, clearly has primacy in attempting to deal with this problem. This results from the delegation of the task of goal attainment, and even the responsibility for deciding what the goals shall be, to the government. One of the basic functions of general election campaigns is to engage people in societal goal definition and the willing of the means for their attainment. It also affords the opportunity to pass judgement on performance in goal-striving by governments. In the United States, a president will be measured by the extent to which he is motivated by identification and adaptation, and by the depth of his commitment to the goals, commonly called the welfare of the Union, and by his willingness to use his office to advance the goals which he thinks desirable.

According to an authoritative American report, *Goals for Americans*, published in 1960, the goals of the USA are for the

> economy to grow at the maximum rate consistent with primary dependence upon free enterprise. ... Technological change should be promoted and encouraged as a powerful force for advancing [the] economy.... The development of the individual and the nation demand that education at every level in every discipline be strengthened. ... Communist aggression and subversion ... threaten all that [the USA] seeks to do both at home and abroad. ... Disarmament should be [the] ultimate goal.

The industrial system tends to ignore or hold unimportant those services of the state which are not closely related to its needs. This includes such services as care of the ill, aged and physically or mentally infirm, the provision of parks and recreation areas, the removal of garbage and refuse, the provision of agreeable public structures, assistance to the impoverished and so forth. The aesthetic dimension, as we have seen, is also beyond the reach of the industrial system, and there is an ever-present danger that members of the system may perceive it as unimportant.

Hence, the cultivation of the aesthetic dimension accords a new and important contemporary role to the state, and if there is any conflict between industrial and aesthetic priorities, it is the state that must assert aesthetic priority against the industrial need. Only the state can defend the landscape against lines of pylons and the proliferation of billboards, and rule that some patterns of consumption, such as motor vehicles in city centres, are inconsistent with aesthetic goals. The state, acting for society, can protect us as we listen and view radio and television from contrived dissonance, or provide alternatives that are exempt from this. It can divert national wealth to remove slag heaps, to beautify highways and to cultivate museums and libraries.

The state can be expected to do better in support of the aesthetic dimension in the future than in the recent past, as we become increasingly aware that aesthetic values demand a much higher degree of public respect, and as society at large creates the necessary surplus wealth to afford the elevation of priorities in this sector.

In the US context, between 33 and 50 per cent of public activity is concerned with national defence and the exploration of space. Western Europeans and developed Communist countries have known similar levels of expenditure on military capabilities. But in almost every society today that balance is changing. Despite the fury of admirals, field- and air-marshals, the checking of Britain's own galloping military budgets and the reduction of commitments East of Suez in recent years has enabled there to be a considerable switch of national resources to other social goals: health, housing, education and regional disequilibrium have been the most urgent foci for these diverted resources.

In a speech on 18 January 1968 to the Council of Europe, the British Minister of Technology suggested that 'there is an inexorable logic in the internationalism that advanced technology imposes. It is the issue confronting all of us in Europe'. It is a vital issue, and one in which Europe lags behind North America and Japan, not so much in new ideas, but in the commercial exploitation of original ideas. In 1963 it was estimated that European firms were:

5 years behind the US in engineering;
10 years behind the US in manufacturing; and
20 years behind the US in marketing,

but that we were catching up quickly. Today, we have still not caught up. Perhaps the contemporary figures are 2, 5, and 3.

Professor Joseph Ben-David, of the Hebrew University at Jerusalem, carried out a survey in 1966 of major industrial innovations. Of the inventions that led to these, ten were initiated by Britain, France and Germany, and nineteen by the USA. Only seven of the European inventions were converted into final products, compared with twenty-two by the USA. The conclusion is that one major strength of US industry lies in its ability to carry an idea through to the final product without a break in the innovative chain. Effective TF would enable European firms to improve their performance, and it should be noted that within Europe, Britain spends more on R & D than any other country – nearly 3 per cent of GNP, or £1,000m. per annum. Britain's technology has been strewn with instances of undercommercialized or irrelevant brilliance since the industrial revolution. Since the turn of the century it has been positively harmful to our social development. TF, and particularly Technological Mission Analysis (see Chapter 2.5), can do much to remedy this tragic situation.

On a comparative basis, the Institut Battelle has developed a model for the EEC, UK, and USA. It is a complete circular model, (as shown in Figure 4) and programmed to be used on a computer.

This model is built up in three phases. First, a socio-economic picture of the future is forecast, including education, population, income, and government spending patterns, along with total expected demand. To this is added an input–output analysis of inter-industry relationships, including use of machinery and materials to meet expected demand. Technological change is specifically catered for by the use of a technical coefficient. The final stage involves a projection of wages, profits and prices, with the introduction of changes in relative prices.

Such a model has the distinct advantage of coherence implicit in

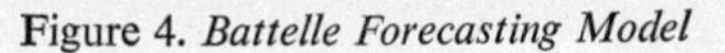

Figure 4. *Battelle Forecasting Model*

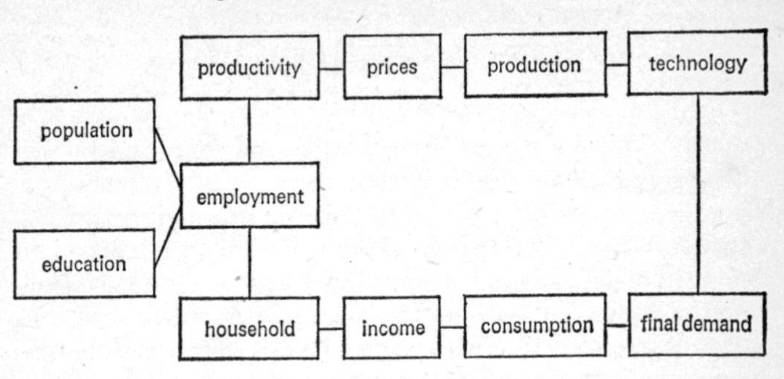

its circularity. Apart from aiding in comparative analysis, the model permits the elaboration of national strategies for the environmental factors which interact with technology to determine the wellbeing of our society. Our discussion has thus returned to our starting point for this chapter. Technology does not work its way alone in society; it is manipulated by members of society whose value-system and goals can be influenced by society if it so wishes. Through TF we have the means to assert social goals and we must look increasingly to the state to speak up in the social interest.

CHAPTER 1.3

The Corporate Impact

In the act of planning, one must reconcile the desirable with the attainable in terms of corporate resources. An arbitrary aim can be worse than none. TF must accordingly be integrated into the overall corporate planning process. TF and planning today exhibit a marked trend towards fuller integration, an integration that may well result in the eventual disappearance of TF as a distinguishable discipline by the 1980s. Such a trend favours a shift from product-oriented towards function-oriented planning.

Recent analyses across a wide range of industries have indicated that the management processes of every technology-based company are particularly sensitive to certain key variables. Observations suggest, moreover, that technology-based companies which recognize and identify these variables, and seek to exploit their implications in the planning process, are more likely to excel in their competitive performance.

The relevant variables can be isolated by asking such questions as:

Is R & D investment consistent with corporate strategy?
Should we invest in the same technologies as our competitors, or in different ones?
How do we maximize the flexibility of our organization structure in the face of rapid technological change?
How can technology transfer best be achieved from R & D to manufacturing and marketing?
What kind of product/market strategy should we follow?
What technical advantages in our products, at what cost, will be needed in the future to give us a substantial competitive advantage?

The need to influence and persuade the customer is an ever-present planning requirement, which is made necessary by extensive use of advanced technology and capital, and also by the related scale and complexity of organization. The relative inflexibility that high investment in R & D and large-scale organization contain, demands and necessitates planning. But uncertainty is always capriciously present, and tasks must be performed so that they are right not for the present, but for that time in the future when all related tasks are completed. Steps must therefore be taken to prevent or neutralize the effect of adverse developments, and to ensure that what is ultimately foreseen actually occurs.

Corporate or long-range planning in technological areas is based on the recognition that any particular technological development may have a variety of possible outcomes; that a problem, or set of problems, will often have a variety of technological solutions, and that technological developments form part of a larger system, thus having to satisfy simultaneous strategies with a system-wide scope.

For successful operations, research, development, engineering and technical programmes can never be considered as things apart from other corporate activities. They must be explicitly integrated into the product, process, market penetration, personnel development, PR and other missions of the firm.

This is but one aspect of the planning concept of coupling. One of the major tasks of TF is to improve the coupling or relationship between successive R & D phases. By anticipating the possible results of each phase, TF can stimulate and guide work to be undertaken in consecutive phases. The integration of the technological forecast into the framework of corporate long-range planning is fundamental to its effective use.

Three degrees of coupling can be distinguished and are indicated in Figure 1. High coupling requires close interaction amongst the technical, manufacturing, and marketing functions of the firm, with multilateral information flows between each area. In contrast, the low coupling case exhibits a one-way, downstream flow of information and, as is the case with moderate coupling, shows no link between R & D and marketing. Those flows existing in the moderate coupling example are of a bilateral nature.

For corporate scientific and technical strategies to be effectively implemented, they must be ultimately reflected in strategic thinking at the programme level, especially in the introduction of new technologies into either operations or the market place. The first step in this process is to anticipate what forms of opposition any given technology is likely to encounter. It may be possible to analyze and consider specific counter-measures and resistances that outside groups will pose.

Figure 1. *Degrees of Coupling in TF*

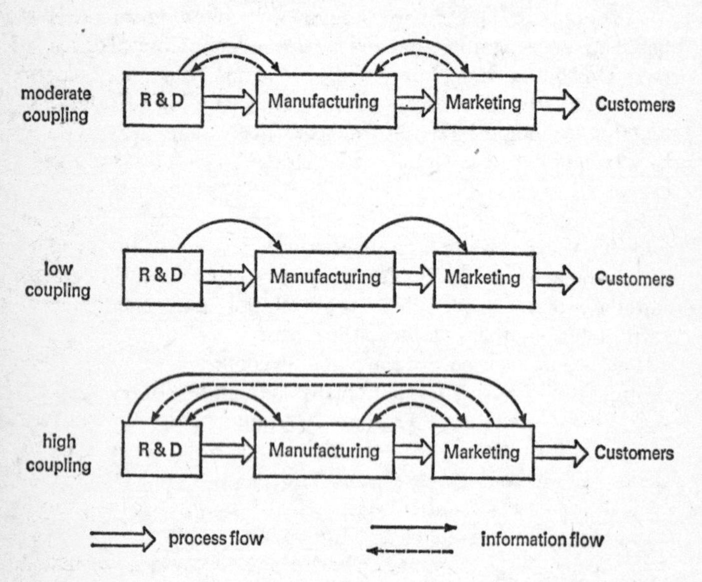

However, the inadequate development of strategy is perhaps the most common and significant factor in industrial-technical planning today. The most typical and serious breakdowns in planning are as follows:

(i) forecasts and plans do not reflect the potential acts and responses of intelligent opponents, such as trades unions, competitors, domestic or foreign governments;

(ii) managers do not allocate limited resources uniquely to exploit the firm's special competencies and offset its particular weaknesses;

(iii) marketing, production, financial and scientific–technical plans are drawn up without careful analysis of the opportunities and threats that technology may pose in the future;

(iv) operating, technical, and service groups' plans are not properly integrated across formal organizational lines. Hence companies can not bring forth their full strengths to bear on critical problems;

(v) simplistic use of rate of return or present value criteria to set investment priorities and to judge operating performance subverts the firm's entire strategic posture; and

(vi) budget allocations, programme plans and performance evaluation systems do not reflect intended strategies; consequently, strategies are never implemented.

TF must be treated as an integral part of corporate planning because a technological element in such planning is necessary to relate strategic aims of the firm to a multitude of technological options. Technological planning itself aims at objectives set in strategic terms, so the two cannot operate effectively in isolation.

Four legitimate strategies can be identified in a TF framework. Briefly these are:

(i) *Leading the mainstream*, or first-to-market. This strategy is based on a strong R & D programme, good technical leadership, and extensive risk-taking. The aim is to produce a steeper gradient of performance over time, and it is an aggressive approach.

(ii) *Drifting with the mainstream*, or following the leader, is based on strong development resources, and an ability to react quickly as the market starts its growth phase. It aims to follow closely the forecast curve, especially in industrial areas characterized by a large number of equally capable competitors. Such an attitude is taken by firms aiming at relatively low-risk profit, and is partly at the roots of present

European complaints about a technological gap between the USA and Europe.

(iii) *Applications engineering*, the third viable strategy, is based on product modifications to fit the needs of particular customer segments in mature markets.

(iv) *Me-too*, finally, is a strategy based on superior manufacturing efficiency and cost control, but an unwillingness to take risks, and an underdeveloped marketing function.

Each of these strategies has, of course, got strong and weak points under different conditions. The first two can scarcely hope to operate effectively without technological forecasts to guide their activities. To respond rapidly to a market leader, companies in category (ii) must maintain a comprehensive range of technological knowledge, particularly where a product life-cycle is extremely short. These points hold true in addition, of course, for the applications engineering and me-too concerns.

The dangers can be overstated, however, of adopting policies

Figure 2. *Gestation Periods for New Products*

	Idea born	*Broad scale launch*	*Years' gestation*
Ban Roll-on deodorant	1948	1955	6
Birdseye Frozen Foods	1908	1923	15
Crest toothpaste	1945	1956	10
Decaf (decaffeinated instant coffee)	1947	1957	10
Gerber baby foods	1927	1928	1
Johnson's liquid shoe polish and applicators	1957	1961	3
Minute Rice	1931	1949	18
Bendix	pre-1939	1953	14
Polaroid Camera (colour pack)	1948	1963	15
Television	1884	1939	55
Zip-fasteners	1883	1913	30
Penicillin	1928	1943	15
Xerox	1935	1950	15

(ii), (iii) or (iv) rather than (i). The overall gestation periods of new products, before they make a meaningful market entry, are a vitally important variable in different businesses. Adler has examined the time which 42 products took to reach the market. They come from consumer-packaged and other goods, and from industrial markets. Some examples demonstrate convincingly that in many markets development time can cover a wide span of an executive's life in a company.

None the less, it is undoubtedly true to say that, during the 1970s, 1980s and beyond, the average gestation period in any particular industry will gradually decrease. Particularly with the advent of technological forecasting one can anticipate a greater appreciation of the alternative applications for ideas.

A second important qualification to the widespread assertion that the market leader and innovator will benefit most from his pioneering activity has been provided by A. C. Nielsen and colleagues in Britain. They examined the success of new products introduced in 89 product classes, covering food, toiletries and household goods. They further analyzed their data by market structure – single brand dominant, leader and runner-up dominant, three or more brands with little peripheral activity and three or more brands with a substantial degree of peripheral activity. They concluded that:

(i) the greatest rewards do indeed go to the pioneers; but that
(ii) significant and frequent successes have been scored by developing a product which caters for a more specialist segment of consumers than the leader;
(iii) that a demonstrable differentiation of a product enables its message to go over clearly;
(iv) that, notwithstanding (ii) and (iii) above, it is unwise to enter a static or declining total market or one where there is a low level of peripheral activity but three or more major brands entrenched.

These two specific examples – gestation periods and market penetration – must modify any tendency we have to feel that the technological leader sweeps all before him. There is yet a further and perhaps more worrying aspect. Technological advancement generates its own exhaustion. Both managements and customers need time to digest, both in terms of acquiring the knowledge and skills to sell and service products and to organize the company's human and capital resources in their support. The computer industry provides an excellent instance where many consumers would be grateful for a sustained lull in development. Already a new generation has been announced before the third generation has been installed, let alone brought on line.

On the other side of the coin a period of rapid technological development can give rise to an inertia resulting from the existence of obsolete but functionally effective equipment, etc. Furthermore, an educational structure without paying sufficient attention to refreshment can fail to meet the needs of industrial leaders as times change. It has been suggested by Sir Solly Zuckerman that Britain has indeed suffered from this inertia, and Hirsch has gone further to suggest that it is an international phenomenon akin to the product life cycle within the company context. This pattern of reasoning suggests that even the most advanced companies and nations are vulnerable in the long-run. To what extent this will be true at a time when organization structures are increasingly built to accommodate a continuous process of innovation it is difficult to say. Certainly, however, one can anticipate that a sudden eruption within a dormant technology will cause industrial and social disruption of considerable magnitude, and in such circumstances a large company will often find it difficult to react as swiftly as a smaller organization might be able.

TF AND THE PLANNING PROCESS

The scope of TF at the policy planning level is the clarification of the scientific–technological elements that determine the future boundary conditions for corporate development. In other words, it is focused on basic scientific–technological potentialities and limitations, as well as their conceivable ultimate outcomes in a wide systems context.

Forecasting and planning at this level face particularly difficult problems, since they deal to a large extent with value dynamics in the society in which the industrial system plays its part. Yet it is at this level that TF can stimulate fundamental research in a significant way, and also provide broad guidelines for it.

A special problem, apart from that of values, relates to future changes in inter-industry structures. If one ignores the acquisition policies of financial holding companies (such as Slater–Walker Securities Ltd), and diversification on the basis of special skills, there remain at least three effects directly related to technological development.

The first is diversification through organic forward and backward integration to form integral functional chains. An example is provided here by the Esso–Nestlé enterprise for the development of single-cell protein on a petroleum basis.

Secondly, there is the invasion effect of one industrial sector by another, on the basis of technological innovation. This is well illustrated by the invasion of textiles by the chemical sector, and invasion of the chemical sector by petroleum companies.

In the third place, a blurring of sectoral boundaries may occur due to technological developments pushed to their extremes. This can be demonstrated by the vanishing distinction between electronic components and systems in an era of integrated circuits.

At the strategic planning level we find the core of TF activity; it is here that the nature and impact of future technological options become discernible in their profiles. Decisions leading to the adoption of specific technological development lines are prepared in the strategic planning phase, and imaginative forecasting is able to influence the course of social and economic macro-processes more effectively.

The scope of TF here is to recognize, and compare by some means of evaluation, the alternative technological options. The focus is not on descriptive forecasts – such as how a new type of machine may look or work – but on assessing feasible systems performance in the light of attainable technological capabilities, and on relating technological options to functional missions. Forecasting must attempt to penetrate into those areas which are to be clarified by operational planning.

At the tactical planning level the task of TF is the probabilistic assessment of future technology transfer. Both vertical and horizontal technology transfer must be considered here, implying that the forecast penetrates beyond the design and performance characteristics of a specific technological system to its utilization in the context of different applications and services, repercussions in the market system, and implications for developments of a social, technological or other nature. Tactical planning will also normally include a contingency approach. For example, the various network planning techniques may be considerably enriched by technological forecasts related to critical obstacles and decision points.

In the face of the corporate planning requirement to determine good strategies by selecting them from as broad a range of alternatives as possible, TF, especially at the strategic planning level, is called upon to enrich this basis for selection. TF must thus be viewed not as an external contribution to planning, but as an integral part of it. To the extent that a technological forecast influences the design of the plan, this plan will also influence technological forecasts. This, of course, is irreconcilable with any notion of TF as some source of absolute or determinate truth. It is, nevertheless, of the utmost importance to understand that both technological forecasts and plans are adaptive to each other, and to make use of this flexibility in corporate planning.

Figure 3 attempts to provide a scheme for the complete TF cycle, applied here simultaneously to all domains of corporate planning with technological implications. It shows the feedback cycles between the environment and the science/technology base at each corporate planning level, and also the triple feedback loop between TF at the three corporate planning levels – policy, strategic and tactical. Dotted lines indicate the possibility and desirability of high coupling.

By showing technological planning in a corporate framework in this way, planning at each level is seen to have a system-wide scope: anticipations, functions and technological systems respectively. The characteristic inputs of TF are indicated for every level, with those on the environment side enriched by inputs from non-technological forecasting.

Figure 3. *The Technological Forecasting Cycle*

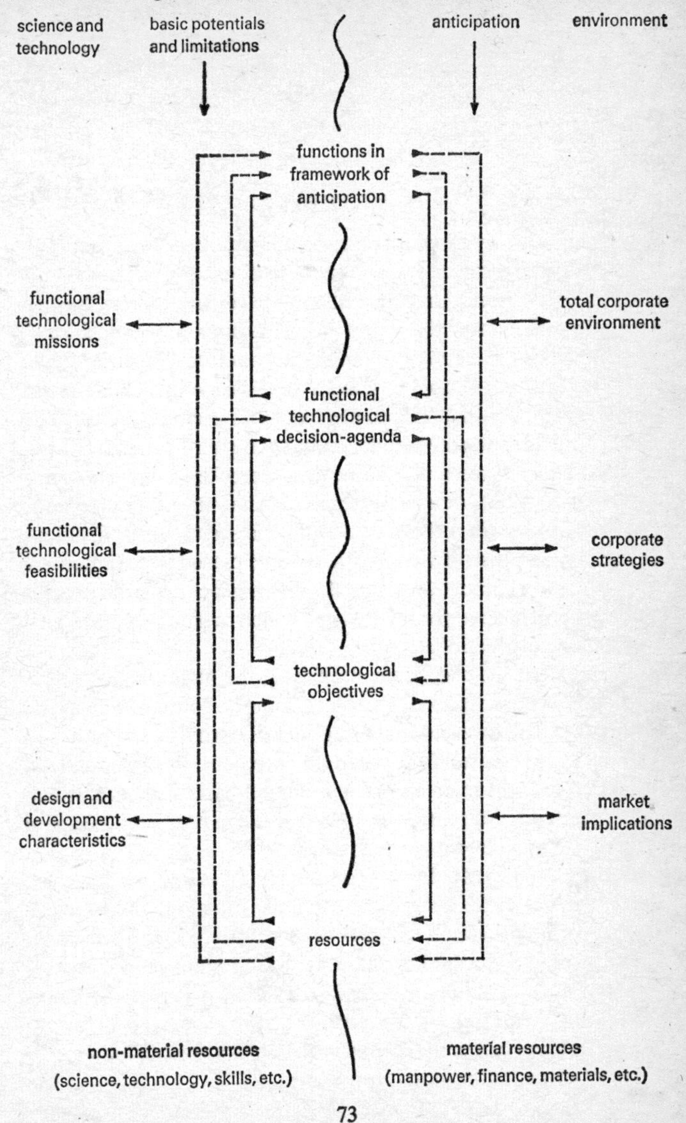

science and technology
basic potentials and limitations
anticipation
environment
functions in framework of anticipation
functional technological missions
total corporate environment
functional technological decision-agenda
functional technological feasibilities
corporate strategies
technological objectives
design and development characteristics
market implications
resources
non-material resources (science, technology, skills, etc.)
material resources (manpower, finance, materials, etc.)

In the context of shaping the future, integrated forecasting and planning should not be allowed to anticipate or bias decision-making, but, as we have seen, should provide the insight into the necessities and consequences of alternative decisions.

ORGANIZING TO MEET FORECAST TECHNOLOGY

In their classic study of the management of innovation, undertaken in 1961, Burns and Stalker were able to show convincingly that certain styles of relationship and job specifications could cater more adequately for the rapid adjustments of organizational activity inherent in a rapidly changing technology like electronics. What they termed the organic system was able to adjust and absorb new roles in a way that a more mechanistic system could not. The latter's natural tendency was to reject such roles. Much has been written, and such research has been undertaken, which demonstrates the vital importance of organizational flexibility if technological advances are to be successfully commercialized. From an organizational point of view, TF should be placed in a framework of all forecasting procedures, thereby making it close to the process of elaboration of a corporate strategy, which must be accompanied by economic evaluations.

Management must concern itself with the technological areas that ought to be developed, the allocation of resources to this end, the extent to which development should be internal or external, and the time-scale required for overall development. In these matters, guidance can come from a systems-oriented basic structure, which offers flexibility, concentration towards the market, and the space necessary for thoroughbred innovators.

In practice TF is usually incorporated in the corporate structure as a refinement of the long-range planning function, and is generally more closely related to long-range planning than to R & D. Indeed, it is often located much closer to top management than is research. A closer alliance with R & D can, however, be confidently anticipated in the years ahead.

The TF input can come from a variety of sources including:

(i) a permanent in-house function;

(ii) an in-house task force (maybe with the help of consultants);
(iii) a *think group*; and
(iv) consultancies contracted to do the job for the firm.

The last mentioned has the disadvantage that external forecasts are of little value to the firm if it has no corresponding in-house function capable of absorbing and adapting them. Considering TF as an in-house function, it is generally found in one of four positions in the corporate structure. These are: as a corporate staff function; as a top-level committee function; as a research staff function; and as an operating division staff function. In the North American context, the favoured position is as a corporate staff function, with three to six persons engaged full-time in TF; but in most large US firms, the TF function represents an interaction between two or more of the four positions specified. The best results appear to come when specialist technological forecasters interact with executives, and when the TF team is composed of both scientific/technical and economic personnel. In line with this it follows that TF should be so organized that it is brought into close contact and cooperation with other staff functions – especially marketing and product planning.

In the corporate planning department of Esso, the TF function assumes what is virtually an educational role for the entire company, with the following groups all participating in TF: economic forecasting, business appraisal, science and technology, energy forecasting, mathematics, and OR. A move from horizontal to product-oriented vertical organization has been followed, with the introduction of long-range planning and TF, by a horizontal control of a function-oriented hierarchic structure. This has arisen because product-oriented structures have tended to act as an obstacle to innovation, whilst function-oriented structures are seen to permit planning aligned with long-range objectives and social goals.

Since both TF and long-range planning favour this function-oriented organizational structure, the logical tendency has been their merging, with the end result that forecasting ceases to be a distinguishable corporate discipline. Such a merger is likely to be accompanied by a radical change in the organizational pattern. The need has been discussed for a closer relationship between marketing and technology, which must take place within a dynamic planning setting, thereby permitting research effort to be directed towards developments that consolidate and reinforce the firm's strengths and weaknesses. Thus the firm must be organized so as to take advantage of its opportunities. This demands good communications, and one way of ensuring this is to have technological and marketing staff reporting to the same director, who should carry responsibility for the future of the firm. We have already described the pattern which emerges in organizational chart terms (see Introduction, Figure 3).

However, successful and creative management can be attained only partially through formal structures – the spirit of management is more important. A fully integrated TF and planning function within the organization requires entrepreneurial qualities at each level, and thus tends to depend on self-motivation in creative people at all levels. Such self-motivation becomes fully effective only where the corporate objectives parallel supreme social and individual goals. It can happen that knowledge of top management plans and policies for the future may not be afforded to the technological forecasters, thus preventing their working in the light of long-range corporate objectives. If all but the highest ranks are excluded from seeing pertinent information, the future can only be forecast as either an extension of the present, or on the basis of pure guesswork.

In concluding these observations on the corporate impact of TF, it is salutory to recall once again the time dimension along which technological change takes place and the competitive threats which the firm must realistically face. The pattern of diffusion of new technology, or indeed of a new idea of any description, both in terms of its operational adoption by producers and its acceptance by users, has a critical significance. This situation should not be confused with our earlier discussion of the gestation period which carries new scientific knowledge to the marketable product stage. By such a time the technological potential to produce colour television sets, for instance, will be beyond doubt. For many manufacturers, with heavy capital investment in previous technologies, the critical question is the speed with which the market for such a product of technology will develop, as well, of course, as the dimensions of its eventual size. One can now see that central heating and dishwashers have diffused much more slowly than washing machines or cars into British households, but the rate at which such diffusion took place, and will in the future continue to do so, is vitally important.

For the follow-the-leader strategy, a careful and continuous monitoring of innovation will in many fields be sufficient indication of a need for concentrated development work to produce a satisfactory competitive product, with sufficient differentiation and even superiority based on exhaustive analysis of market response to the innovator. This monitoring activity, not just of competitive activity, can, of course, extend to the register of new patents and of other listings of technological advance.

There are three major techniques currently used for ensuring that a continuous evaluation of competitive activity is sustained: the life cycle model, technological mapping, and strategic analysis. The life cycle model has been used to ensure that marketing and R & D investment costs are recovered over a relevant period before product sales are undermined by substitutes. The experience in pharmaceuticals, for example, is that within five years substitute drugs have taken over the greater share of any new market opened up. Hence payback tends to be over a three-year period. Such an approach also highlights ways in which competitive seizure of a market can be forestalled by the innovator's own introduction of substitutes.

Technological mapping can be sketched from the basic work already carried out for the technological forecast. A competitor's known strengths in research, registered patents, existing range, etc., will be evaluated in terms of technological objectives thought to be suitable and likely to be pursued. Particularly for the follow-the-leader company such a procedure can enable management to isolate areas of likely differential advantage over competitive offerings.

Strategic analysis is a logical extension of technological mapping.

Valuable indications can often be gained from announcements of mergers and acquisitions, or from new product introductions which indicate a new direction of activity. Competitors who monitor this type of information objectively and systematically can frequently spot where their relative opportunities lie and where they may become vulnerable.

In essence, the final injunction to every business is to avoid introversion in the planning process. It is a warning easily issued but difficult to heed with true effect.

CHAPTER 1.4

The Functional Impact Within Business

TF will not nestle happily within the corporate planning sector of the business. The rigour which it brings to its field of evaluation has ramifications in many of the functional areas of the business. My comments in this chapter will, however, be of a general nature since in Chapter 2.5 we will be examining specific technological missions in detail.

Worthwhile technology is market-directed technology, so it is appropriate at the outset here to pick up again the theme presented in the Introduction. The market directedness must, however, frequently encompass much more than a domestic economic system. The demand on resources is so heavy in such fields as aircraft that home demand cannot alone justify all the development work required – thus we must look to world markets, and frame production to meet that end. Rolls-Royce has for long adopted this stance in the sale of aero-engines and the RB-211 engine was designed and developed quite specifically for the world airbus market. ICL have found the same objectives to be technologically essential in their development of computers in the face of extremely powerful competition.

Our engineering industry, defined to include aircraft, shipbuilding, vehicles and electronics, produces about one third of Britain's manufacturing output, and nearly half of her total exports. It is of paramount importance that marketing and technology act together to compete successfully, for instance, in the North American and European markets. Since technology is today essentially world wide, the success of a technology-based industry must depend on its ability to see worldwide opportunities, to market internationally, and to meet worldwide requirements. Such requirements are difficult to anticipate. Customers respond to market offerings as well as developing specific individual needs. As conventionally understood, marketing thinking fails to indicate the need to adjust to futures further ahead than that which consumers can easily visualize. Although goods and services on the market play a central role in the adaptive process, as it is through them that change is communicated to the firm, such offerings can only change in line with available assets, techniques, resources and experience. Innovation is critically expressed to the world market through product development, and corporate identity is often generated in terms of product function.

To develop in a planned way, a firm must almost certainly have a recognizable and recognized identity – a role understood both by its employees and by its other public. Such an identity, of course, must be capable of adaptation. Given a positive role or identity, managements can view current news, technical innovation, commercial development, social, political and economic change and ask themselves, 'What's in it for us?' Thus, for example, a firm in an industry subject to threats from technological innovation will review its environmental position and opportunities, and decide on its field of major interest. From a technological point of view, new products can result from either the discovery of some property or concept for which demand might be stimulated, or from the identification of some market requirement which could be met by a suitable device.

In either case, the product only has meaning within a market. Thus marketing research may be employed to identify devices which are required and those technological innovations which, although feasible, cannot be economically produced and marketed. Anticipating market demand for a new development is difficult. The results of, say, a Delphic exercise must necessarily be analyzed by the marketing researcher and social psychologist; they can help with modes of questioning and the interpretation of results. An even more hazardous but equally important task is forecasting rates of diffusion for new products. There is little doubt that the increasing attention to technology transfer which is implicit in TF will lead to a reduction of diffusion times for innovations. This time appears to be decreasing at a logarithmic rate of shrinkage. Hence, the premium for success in marketing must increasingly shift to the first-to-market principle which was enunciated in Chapter 1.3.

This requires that the alert firm must plan to be amongst the first to bring out a new product, or break into a new market, since competition thereafter will generally force prices down rapidly, thus depressing profit margins and return on investment. Organizationally the shorter cycle demands overlapping or simultaneous functional planning. Manufacturing, for example, may begin to frame its plans, and marketing may set target dates for product introduction, before R & D planning is complete. Consequently, short-cycle businesses need close coupling between product marketing specialists and technical staff. Marketing managers will tend to be knowledgeable about technology, and may often contribute substantially to product definition and development.

In the case of longer life-cycles, hitherto characteristic of many markets, there has been adequate time to learn about competitive market developments and to plan to counter them. There has been less need for exceptional market sensitivity. In many companies, the emphasis has therefore been on established procedure and routine. As a result, planning was sequential, with detailed R & D plans being completed well before manufacturing and marketing planning were initiated. Marketing and manufacturing were seldom deeply involved in technical planning, and marketing managers often unfamiliar with the specific technical problems or objectives. The marketing group tended to be volume-oriented rather than response-oriented, since new technical problems were rare and the coupling between marketing and technical staff was low.

From a general viewpoint, the advent of TF enables a company to foresee or predict more accurately the slope and duration of the product life-cycle, thus increasing the benefits to be gained by use of the concept. In the past, the dimensions of the product life-cycle have had to be estimated simply through experience, but TF can now predict these values with greater speed and accuracy. The example found in the pharmaceutical industry, in which substitute drugs now normally take over the greater share of any new market opened up within a period of five years, has already been cited.

TF in relation to a product's life-cycle will not only help to ensure that a more rational approach is brought to product planning and marketing, but will also assist in creating valuable lead-time for important strategic and tactical moves after the product is brought to the market. Specifically, it can facilitate the development of an orderly series of competitive moves in extending the life of the product and in purposefully phasing out dying and costly old products.

A further deep seated transformation of marketing's role can also be envisaged, arising from the effect on the customer of an ever-increasing rate of technological innovation. The customer's buying decisions will become more difficult; he will be confused by the increasing complexity of the goods and services he buys – especially those recently introduced; and he will be irritated by his inability to understand the nature of such new goods and services – thus tending to resist using them for a substantial period after their introduction.

The importance of this latter effect on the sales of a rapidly advancing technology-based company is obvious. Resistance may keep sales low for some time, thereby placing in jeopardy a policy of maximum technological advance. Unless action is taken to combat this, an increasing rate of innovation would suggest that resistance to purchase will rise.

A new importance must accordingly be ascribed to promotional techniques in terms of the lowering of such resistance. An essentially long-term educational effort will be needed to help consumers understand the nature of new products in order that they may readily accept them. With an accelerating pace of innovation, this educational task becomes larger, and hence requires more time for the percolation of information. Educational effort will require an ever-lengthening lead-time for the products with which it is dealing. This in turn will necessitate a new sales approach, a didactic orientation in selling. Salesmen may well become teachers, with marketing's task to provide an information service for all customers, actual and potential.

Much marketing effort will no longer be concerned with existing products, but with describing the qualities of future products even before these qualities can be precisely known. In order to fulfil such a role effectively, a clear idea of the nature of future products must be gained well before they are to be introduced. Marketing must inevitably be closely involved in the development of TF and R & D planning to achieve this.

Marketing will also be scanning distant geographical markets as sources of threats and opportunities. The increasing speed of transportation now allows competitors to invade markets with greater timeliness and responsiveness to customers' needs. In addition, higher volume of shipments will support new packaging and handling systems, and in some cases specialized one-product transportation techniques such as pipelines.

This demands an improved conceptualization of the distribution process; the integration of handling, packaging and transportation systems; plus the technological advances in these systems – all of which will lower costs so that goods can travel more miles for the same outlay. Distance will become less and less a barrier to competition. Competitive challenges will also come from non-traditional and unexpected fields.

Sophistication in the product offered in the market place also has its compensations. It affords substantial scope for product superiority and competitive improvements to a business. The facility to improve delivered capability for the customer is always present. Above all else, however, TF re-establishes the necessary balance between R & D and the market place as legitimate sources of innovative ideas in the business and integrates this, the realistic state of affairs, into the planning process of the business.

FINANCIAL ASPECTS

Wherever a high value of investment is required, and this is particularly so in the development of emerging technologies, failure to persuade customers to buy or the mis-timing of innovation will prove extremely expensive. The pursuit of technological red herrings is also, of course, an inherent risk in such activities. Against such a background, it will be readily seen that TF can impart greater coherence to the financial analyses which must necessarily be undertaken. Furthermore, if, as is often the case, the state plays a more active role, the cost and associated risk to a business can be further reduced. Suitable justification for support to indispensable industries such as supersonic travel is readily found in the requirement of national security and the needs of national prestige. Modern technology thus defines a growing function of the modern state, which can be expected to result in a somewhat modified technical/financial framework for developing industries.

At the corporate level, the present trend towards greater emphasis on financial management, or management accounting as opposed to traditional financial *accounting*, can be expected to continue and even to accelerate. Such a revised financial role will pay minimal attention to pure stewardship reporting by simple compliance with legislative requirements. A new approach will have to be taken to risk-taking. This will probably not involve eradication of the present excessive conservatism in the training of accountants, but will quite definitely require the application of improved techniques of risk-evaluation to decisions with financial aspects. The alternative is for traditionalists in the finance area to hold back worthwhile ventures on account of delayed paybacks, and so forth.

The adoption of TF will both demand and make possible a more realistic approach to depreciation policies. Since depreciation can result from a variety of causes including technological obsolescence of equipment or products, the accountant must incorporate these phenomena into his normal concept of depreciation involving the spreading of costs over the expected physical life of an asset. The fact that an asset has not been fully depreciated in the books of accounts can hardly be advanced as a valid reason to delay innovation decision-making, save in the most exceptional market conditions of near monopoly. A forward-looking, progressive view must be taken by accountants, in line with the need for the rapid application of new science and technology to British industrial production.

Carter and Williams have studied these problems of investment as presented by basic research, the communication of results of research, the supply of trained personnel, the stimulation received from other firms, and not least the supply of money and the effects of taxation. They followed up with a further examination of those features critically affecting the plans of industrialists when investing money in new plant and equipment, and in research.

The large firm may well have the funds available to put on a crash programme of, for instance, R & D, but high investment ratios in technical development were found to have four particularly significant implications for management:

(i) they require a serious and continuous evaluation of technology procurement alternatives: whether to buy technology through licensing or through hiring consultants; whether to buy a company in order to acquire the latest technology in an unfamiliar field; whether to hire top people with the specific technical competence desired; or whether to develop additional technical competence by internal training in order to stay competitive;

(ii) they usually accelerate product and process change, which in turn requires an adaptive organization, which can quickly shift to new levels of efficiency and effectiveness as technology changes the work to be done;

(iii) they usually mean a dynamic product market, which implies three requirements: clear visibility of resources, thus permitting management to cut off a development project quickly or to switch resources into a new technology; explicit strategy formulation permitting a clear definition of project alternatives enabling management to choose more wisely amongst them; and a well-developed planning system to permit the firm to redirect its resources promptly and effectively;

(iv) they require closer supervision of technical efforts, since the firm is highly dependent on technology for competitive survival and therefore commits proportionately more resources to the effort. As a result, senior management must know more about technical problems and performance.

Any financial function, in the light of these remarks, must exhibit flexibility, adaptability in resource allocation, effective control, and the ability to evaluate alternatives over the long term. The potential contribution of TF rigour is not hard to discern.

In forecasting the future uncertainty is an unavoidable problem; probabilities enable us to make sense of it. This implies risk which is difficult to calculate; but the difficulties do not mean attempts at mastery should not be made. Error can be reduced in deciding which of various alternative approaches to development should be adopted by appropriate application of discounted cash flow (DCF) techniques.

Figure 1. *Option Diagram for Development*

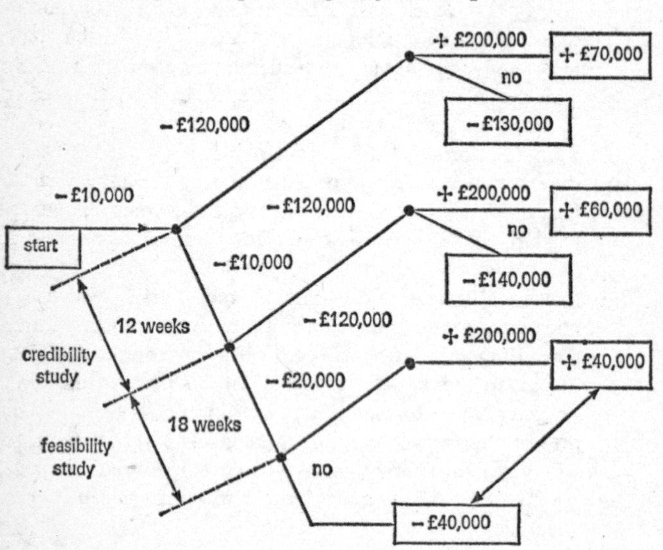

In Figure 1 the introduction of a new capital project is illustrated, on which £10,000 has been spent in preliminary studies. At the first node, the option exists for carrying out further studies, requiring twelve weeks to complete, at a cost of £10,000. Alternatively, the decision could be taken to embark on serious technical development and production planning, at a cost of £120,000.

A second course of action becomes available after the latter choice has been taken: either drop the project and lose £130,000 or sell and install the plant for £200,000. In the latter case the net cash flow will be £70,000.

If further studies were undertaken – such as a feasibility study over 18 weeks at a cost of £20,000 – it may be found that a position is eventually reached when the net cash flow from successful sales is at a level where neither option is attractive. At such a point, the decision to proceed with or abandon the project cannot be made, since it is too late for it to be effective.

Such information aids the accountant in business in dealing with technological change and development. An additional area of concern, however, has arisen. In certain industries it is the cost of *information*, rather than the cost of development, production or distribution, which has become the single most important factor governing the profitability of projects or new ventures. Such information is gathered, largely to reduce risk. Consequently, cost/benefit criteria should be amongst the accountant's tools so that he can assist in ascertaining that level of risk at which a decision should be made. Nevertheless, with the increasing adoption of TF, much relevant data is becoming more widely available, and although this tends to decrease its value in competitive terms to any one firm, it also reduces its cost. The government and in particular the new Department of Trade and Industry see this as an important part of their task in stimulating development.

In these comments on financial aspects there has been the general assumption that the supply of money is not a major problem. This is certainly true of many of the larger business concerns; but it must be remembered that much technological development comes from firms which are less than large, certainly at the time of its inception. In this context, TF can once again play a significant role in providing a coherent basis for funding. The technological forecast will soon come to be expected of the finance director as he negotiates for funds, in precisely the same way that a marketing research study became an essential weapon a decade or more ago.

One final comment on funding is appropriate. Whilst the stock market has traditionally favoured growth companies, it has hitherto had little opportunity for making realistic assessments of growth futures except by the analysis of past performance. TF can be expected to play an increasing role in this field in the securing of better stock-market valuations and hence a greater encouragement to development and growth in the well-planned sectors of industry.

RESEARCH AND DEVELOPMENT ASPECTS

R & D planning and budgeting has traditionally been an area of uncertainty comparable in magnitude with the advertising sector of the business, and often involving much greater resource commitments. That uncertainty does not so much surround the technical feasibility of achieving given goals as the capacity to meet them in an economically viable manner. Furthermore, the means for their attainment are often problematic. T F within the R & D function area of a business accordingly demands close attention to the formulation of corporate goals and the detailed examination of the process of technology transfer. Specifically, R & D planning must:

- (i) establish meaningful research objectives;
- (ii) ensure the organization is attuned to the firm's major technological threats and opportunities;
- (iii) develop an overall strategy into which research is integrated;
- (iv) develop an overall strategy which evaluates research projects in the light of goals and capacities; and
- (v) organize research and operations for a maximum transfer of technology from research to operation.

For any particular firm, R & D policy necessarily depends on its resources and on the sector of industry in which it operates. Studies have shown that many firms have no defined policy for R & D, even though they do have R & D departments. Obviously, R & D should only be undertaken with a clear idea of how and what it is to contribute. T F can give direction to such efforts. For example aircraft companies stand or fall by their R & D endeavours, and this is integrated with marketing research and future commercial plans in a technological framework.

It is suggested that management recognizes only two ways of

Figure 2. *Fundamental Research Planning Cycle*

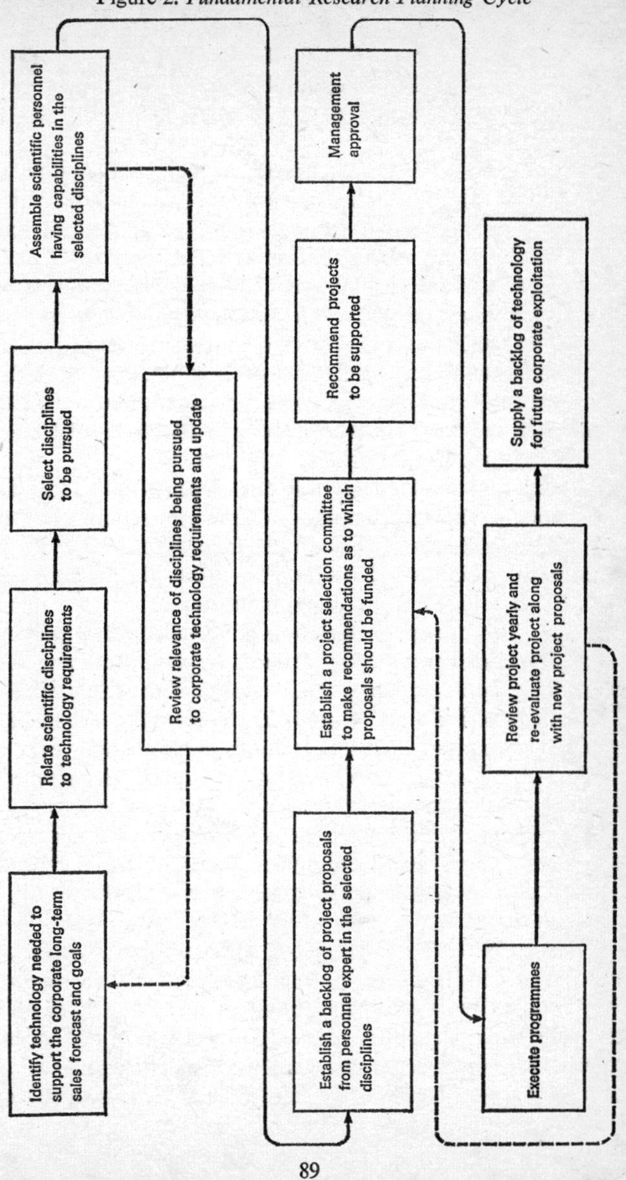

organizing R & D, namely free and fundamental, with unpredictable results; or strictly applied to objectives formulated by men who may not have scientific training, but have many problems to solve.

In either case, R & D effort is somewhat unsatisfactory. If research is too strictly applied, innovations become little better than developments, and if research is too unconstrained and fundamental it is too expensive to justify. The results of R & D must be translated into profitable or viably subsidized products, or more efficient means of production to ensure continuing corporate success, and a judicious combination of the two approaches makes more sense than total espousal of one or the other. Figures 2 and 3 outline typical R & D planning cycles for fundamental and offensive/defensive research efforts suggested by Blackman of United Aircraft Research Laboratories.

The successful development of R & D depends on the transfer of technology, and the complexity of the interactions between changing technologies. Furthermore, innovation, especially in more traditional industries, is very often effected by the introduction of technologies from other industries. There are many examples of functions performed by products and processes in one technology being competitively replaced by those from another, often with a significant increase in the value of important performance characteristics giving a distinct competitive advantage; e.g., the replacement of kerosene lamps by electric light bulbs.

It follows that any review of present and intended development should pay special attention to every component, process, product and technology, outlining alternative ways of performing the same function, and indeed adding new functions. Developments in one technology may of course create opportunities for development in others.

Relevance analysis is a specific T F technique which can be fully employed in planning complex R & D programmes in support of wider objectives. Initially the objectives must be specified on a time scale, and each activity necessary to secure their attainment described in as much detail as is possible, considering both known technological systems and those deemed likely to be required for

Figure 3. *Planning Cycle for Offensive and Defensive Research*

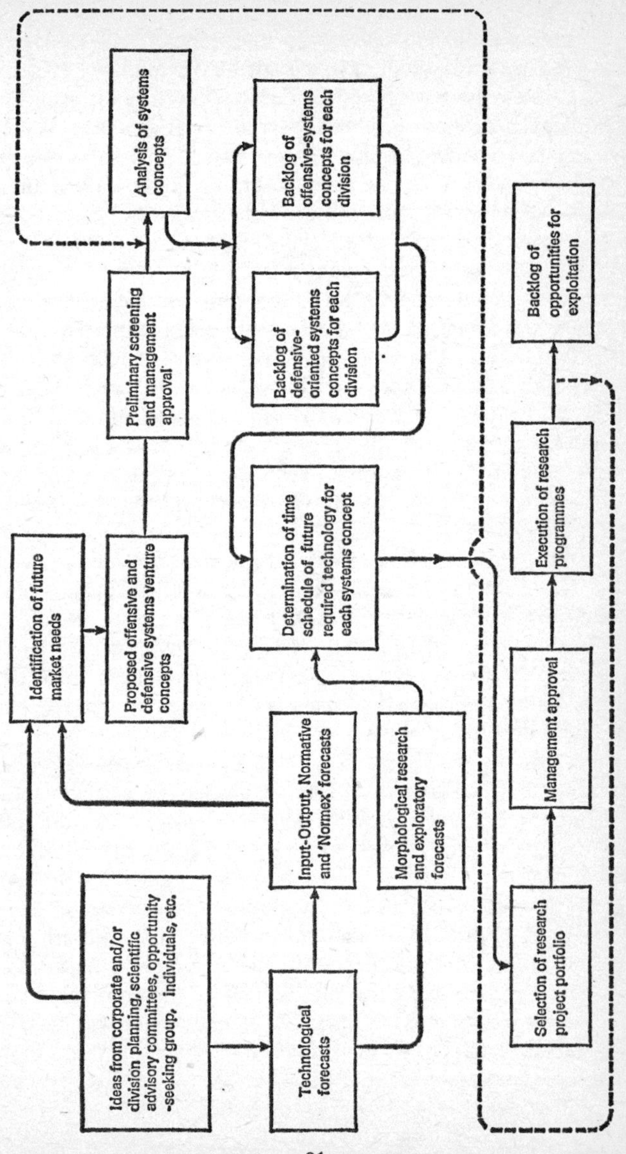

the success of individual operations or activities. (Chapter 2.3 discusses relevance tree analysis in considerable detail.)

Each system that cannot be provided for with currently available technology must then be defined in terms of performance characteristics, and assessed in quantitative terms. This gap between what is currently known and what must be found out is the technological deficiency. Priorities must then be attached to each of the recognized deficiencies and decisions made about approaches to adopt in their attainment, such as the carrying out of research internally, or contracting it out to other bodies. Available skills, facilities, workloads, costs, and current research approaches in progress must each be considered. Some aspects of a relevant technology may be known but protected by patents. In such instances, much effort can be avoided by a licensing arrangement rather than by attempting to reach the desired state of knowledge entirely through the firm's own efforts. Each deficiency must be met in a sequence depending upon its relevance and critically for the total effort.

The next stage of the analysis is to generate specific R & D proposals directed to meeting deficiencies of high priority not already covered. The whole R & D programme may be scheduled in relation to revealed priorities, the difficulties appraised (so far as they can be envisaged), and the required resources specified. Figure 4 indicates the various aspects of technology transfer that aid in this analysis.

Such a systematic approach to R & D allows an evaluation of the extent to which different efforts are directed towards similar or related ends. Particular projects, unimportant in their own right, may thus be selected because of their relevance to overall objectives.

Communications must play a vital role in the success of interrelated R & D projects, in the same way that communication is vital between R & D and marketing. It is uneconomic for groups to work in ignorance of either one another, or their mutual goals. When a research proposal is presented in functional terms, R & D personnel must in no way be restricted in the approach to adopt in solving problems referred to them.

Figure 4. *The Transfer of Technology*

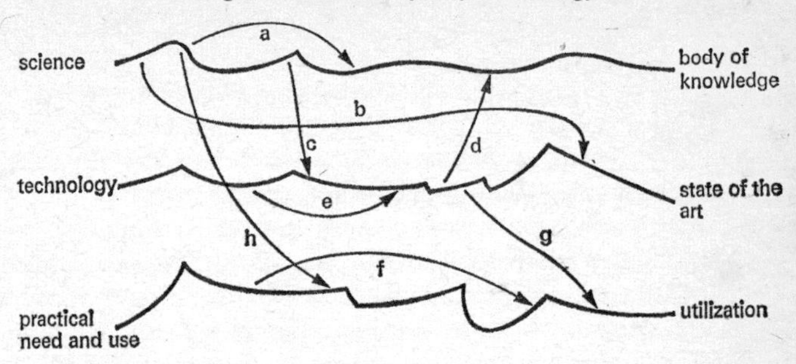

(a) science to science
(b) science to technology (slow)
(c) science to technology (fast gap filling)
(d) technology to science (e.g. instruments)
(e) technology to technology
(f) use to use (diffusion)
(g) technology to use
(h) science to use

How close a company is to an understanding of the contemporary state of knowledge has important implications for management planning and decision-making. These implications may be considered under three headings.

Stability is a function of the distance from the state of knowledge. A firm working near the state-of-knowledge boundary must keep trying for rapid advances like those through which it achieved its current position. Its market position is perpetually in jeopardy from all competitors working in the same technical area.

For firms well back from the boundary, radical breakthroughs are unlikely. Technical progress is evolutionary, with little innovation. Breakthroughs by immediate competitors are no serious threat; the danger is that breakthroughs in other industries and other technologies may make the entire mature technology obsolete.

Predictability is low for firms near the state-of-knowledge boundary. Since their researchers are working in areas of partial knowledge, the nature and, even more, the timing of results are difficult to foresee.

Conversely, far from the boundary of the state of knowledge, where breakthroughs are unlikely, predictability is high. Specific small improvements in products or processes can be foretold with confidence and timed with a high degree of accuracy.

Precedent, which underlies so much management activity, is sparse near the state-of-knowledge boundary. Past experience supplies little guidance to help managers to judge whether people are doing a sound job, whether capital should be committed to a particular project, whether the product has a commercial life – whether, in fact, the entire effort will be profitable. Their task, it must be recognized, differs greatly from that of their counterparts in firms far from the state of knowledge, who can rely on established management doctrine that prescribes the scope of managerial discretion.

Rapid change in the state of knowledge means rapid obsolescence of managerial decisions. Planning assumptions are more quickly superseded by events. Since even the most carefully made capital investment decision may turn out badly, rapid payback of investment or flexibility in capital facilities, or both, become crucial. Opinion differs as to whether or not research and development time may be bought. Without doubt, R & D activities must be initiated in time to be commercially effective, but some do not believe that firms can necessarily buy back much or even any lost time – no matter how extensive their cash resources – to reach particular goals.

Creative science is, of course, in many ways distinct from the diffusion of economically viable science, but when the latter constrains the former, as it often does, the outcome is that the innovation that science should foster suffers. This is more likely to happen with increasing concentration and economies of scale. Big science, big technology and the art of managing large systems are ever becoming increasingly important. Yet the individual researcher, the lone inventor and the smaller firm grow in parallel significance. (The true tragedy seems to await the medium-sized business.) Such a pluralistic system is as essential to science as to the economy if the best use is to be made of opportunities and circumstances.

New ideas often come up against conservative attitudes in large-scale industry, so the small firm will perhaps always find champions in many sectors. The Industrial and Commercial Finance Corporation, (ICFC), suggests that the small company sector may be more profitable and efficient than the large, and that, even in apparently high-risk, expensive technical innovation, the small firm may have substantial advantages. The recipients of the Queen's Award to Industry have included several small firms whose success is based on technical innovation. Small company technology is not only cheap in macro-economic terms; sometimes it is the only possible method to begin when the market is small. The initial development of lasers in Britain was pioneered by smaller firms.

In conclusion, it can fairly be stated that TF, including a strong normative component, will increasingly determine the nature and the growth in volume of fundamental research. The latter, in turn, will supply answers to questions concerning ultimate potentials and limitations that TF will put to fundamental research. Furthermore, the attitudes and techniques of TF, especially relevance analysis procedures for normative forecasts, are applicable to the stimulation and guidance of fundamental research contributing to social goals. This process will also provide spur and guidance to the technology transfer pattern.

TF is the most effective available means of filling the gap to maintain continuous fast growth, and it will strongly influence the pattern of vertical technology transfer, especially by greatly improving the systematic exploitation and development of complex technological systems, here most typically by a morphological approach and extrapolation. The horizons of application and service engineering will be widened considerably by TF, and the trend towards greater emphasis on horizontal technology transfer will also be strengthened thereby.

The benefits of TF in R & D planning have been simply summarized by Honeywell, which began using it in 1967.

(i) Divisions are better informed about research programmes, which has meant that they have given more realistic thought to their technological needs and developed a greater respect for the potential contribution of R & D.

(ii) Both sides have come to a clearer understanding of the crucial importance of coupling between R & D, production and marketing.

(iii) The company's R & D programmes are now better balanced and less problem oriented. Now a clear indication of balance between fundamental and offensive/defensive work has been achieved.

PRODUCTIVE ASPECTS

The future undoubtedly holds productivity gains in store for manufacturing production and superior design. The costs of many materials, products and services will also continue to decline. The industrial designer is increasingly cognizant of relevant new developments, of new possibilities in materials, transportation modes, the aesthetic and the health factors affecting the quality of our life. Designers are joining planners, managers, scientists and technologists in building a future which is humanly habitable. Every planner is a designer – he designs the future. The designer must not only consider the particular task with which he is concerned, but also the context within which the design must function and be judged. His control is over the former, not the latter. Any design must be fully compatible with production techniques that are available or can readily be brought into being within the planned production time-scale. The designer must inspire sufficient confidence to enable people to act on the basis of his forecasts. The technological design forecast is, however, only a starting point – not an end in itself. Given the forecast, one must ask what is to be done in the light of it.

In their traditional markets there is very little security for raw materials. Suppliers and processors of raw materials can expect competition between materials to be increasingly severe. Simultaneously, new markets for traditional materials are opening up as technology enables characteristics of the materials to be modified for new purposes or as it provides ingenious combinations with other materials to serve special needs.

To exploit or protect material markets requires time and money. Although many businesses may be able to generate the necessary funds, they can seldom buy research time. Furthermore, it is usually not enough to come up with a new material. These new materials frequently require new production processes and equipment. In many cases this means an equivalent research effort in manufacturing, and large investments in new capital equipment.

The technological obsolescence of existing materials, production systems and fabricating techniques also brings changes into the selling of materials. There will be far more selling of technical qualities of materials, rather than selling based just on delivery or price. The selling effort must and will become far more technical, because material salesmen will need to translate customers' special needs into product opportunities and applications.

Maintenance facilities – the provision of spares and service – become increasingly important as trade develops in advanced technological products. This cost of maintenance increases as a proportion of operating costs with advanced machines, and can even be the deciding factor between two alternative machines. Such increasing technological sophistication makes it necessary to guarantee that equipment will meet customer requirements. This is best done, on an international level, by an independent body which endorses the quality assurances of manufacturers. For instance, the British Calibration Service has recently been established for this purpose, and similar authorities are being established in many fields.

The relationship between productivity and improved technology is not new. Some more obvious changes that will ever increasingly affect productivity are:

(i) greater use of low-cost automation devices, reducing the skills required by operators and increasing the number of machines under one person's control;

(ii) greater use of numerically controlled machines in all aspects of machining;

(iii) application of process control techniques over a wide area of manufacturing and production;
(iv) automated assembly, with groups of standard automated assembly machines replacing human manipulation in high-throughput repetitive production work;
(v) improvement in quality of product with growing efficiency of automation, and feedback influencing the traditional patterns of manufacturing inspection;
(vi) transport systems – involving containerization, standard palletization, packaging, etc., linked with harmonized methods of mechanical handling and automatic warehousing; and
(vii) steady growth in the use of teaching machines and audio-visual aids as means of instructing and retraining operators, and as means of communication.

The effects that may be expected from these productivity changes are on the pattern of work; on the need to break down the boundaries between the various disciplines, with implications for processes and techniques; on the increasing sophistication and integration of production processes, rendering them more vulnerable to blockage at any point; and on the overall rate of flow of technology which will, along with transferability, make it increasingly difficult to maintain a technological lead.

At the periphery of production technology is the computer-controlled machining system – almost entirely a British development, and far ahead of anything else in the world. Indeed, it is the biggest advance in production technology thinking made during the last half century. The system is an extension of numerical control, which was a significant development itself. It requires no jigs or tools, but uses instead a 'parts programme' and a tape. It reduces scrap dramatically due to fewer mistakes than is the case with human control; it ensures (via a continuous flow of control signals) that metal is cut for 40 per cent of total shift time as opposed to the 20–25 per cent that the manual operator's machine tool is cutting metal; and enhances the reliability, accuracy, and speed with which complex, three-dimensional shapes are machined.

TF is a tool for the identification of such changes and their integrated development along a wide front, thereby ensuring that such development is phased in and planned for in a coordinated way. Such an impact in the productive area of the business can be simply stated. Its ramifications are far reaching when the cost implications of not identifying technological changes are so great.

PERSONNEL ASPECTS

It is evident that the harnessing of technological innovation requires specialized manpower. Organized knowledge can be brought to bear, not surprisingly, only by those who possess it.

However, technology does not make the only claim on manpower; planning also requires a comparatively high level of specialized talent. To foresee the future in all its dimensions, and to design the appropriate action, does not necessarily require high scientific qualification. But it does require the ability to organize and employ information, or the capacity to intuitively draw upon relevant experience and forecasting methods. In addition, specialized talents must work in combination. This united specialization is only possible if the specialists are brought together by a common and accepted set of goals. This holds true from the standpoint of the employee no matter what his society's official ideology is, or whether he happens to be employed by a government or commercial corporation.

Technology, in the form of highly automatic equipment, points to the fact that many skilled-operator jobs will be either eliminated or simplified. Similarly, many unskilled jobs may also be mechanized out of existence. To the extent that the equipment content is greatly increased or is unique and novel, a larger and more skilful maintenance force may be required. Thus, to take all these factors together, the net effect on the work force is not generally predictable, other than to say that in some instances it will be severe and in others trifling.

A further important outcome from advanced technology is the development of highly complex once-off or very small batch products, such as space vehicles, missiles and certain computers. They require an extraordinary amount of skilled assembly, testing and inspection. This raises the technical content and level of the work force to a very significant degree. It would seem unlikely that any operator, skilled or otherwise, will be able to remain in one class of activity throughout his working life. The character of his work will change, and no doubt extend rapidly during that working life, and problems of instructional boundaries, vocational training and retraining, will be critical.

To take full advantage of technological change, training cannot be confined to the operative level. Marketers and particularly salesmen in technological industries will require detailed technical and engineering knowledge, to afford the necessarily higher coupling as well as guaranteeing better customer service. R & D staff, engineers and other technical personnel active in industry are at present given insufficient quantitative and qualitative opportunities for the maintenance and renewal of their professional skills. As a result, they may not be able to grasp technological opportunities, and they may not be certain how to utilize their younger colleagues in the most effective way. This must be modified since human resources form the most important technological asset of any firm. Both traditional training systems and established reward systems of firms will have to be reviewed, along with their organizational structures, in order to encourage sound appraisal of, and response to, technological opportunities.

For the trade unions, changes from technological sources will mean:

- (i) progressively less manual skills;
- (ii) progressively greater competence, understanding and ability to operate increasingly sophisticated machines;
- (iii) fast growth in individual responsibility for increasingly more valuable machines; and
- (iv) increasing the skill of the entire working population from the managing director to the apprentice.

TF alone can enable sufficient advance warning of impending change, thus strengthening the prospect of common aims and aiding in the bargaining process. If TF is not employed, the firm may find itself without the right types of skills to take advantage of developments, and this will have repercussions on all aspects of the firm's operations. With TF the time-scale and the requirements for manpower planning and organizational development are made available formally and with greater certainty. TF holds out the prospect of reduced social conflict within the organization by building into the planning process factors hitherto considered to be totally outside control.

This concluding comment, on human resources planning, echoes our discussion on financial, R & D and productive aspects of the business. We have demonstrated how TF makes possible *a more coherent approach to planning within all functions of the business.*

It has been the deliberate intention in Part One to confine comments to the managerial implications of TF, without any description of the techniques, since I am firmly of the opinion that few of the major potential points of impact where TF will have substantial influence are yet aware of its potential ramifications. My comments have concentrated on technological change rather than the techniques themselves, since it is the phenomena of change that TF can potentially tame – not completely but sufficiently to give many facets of the modern business and social planning a new, urgently required, coherence. That coherence can work for the good of the individual, the firm and society at large, and need not necessarily be used as a tool for the more effective manipulation of all three by a small technostructure.

PART TWO

TF – THE FORECASTING ART

In Part Two we examine the techniques of TF, the forecasting art. Exhibit 2 provides a sequential description of a commonly employed pattern of TF utilization in business. It will be seen as we have discussed in Part One, that the forecasting process commences with the identification of company objectives. Provision is made for a feedback loop to secure the modification of those objectives if appropriate.

Two major avenues of forecasting are available: extrapolative forecasts in terms of extant technological know-how; and normative forecasts. Two twilight categories are to be seen in terms of morphological analysis and normex reconciliation process (*norm*ative *ex*trapolative). Extrapolative techniques afford their own estimate of time-scales on the basis of resource-allocation assumptions. All other approaches require time-scale forecasts and the Delphi technique is most widely employed.

Each major technique of forecasting and the overall pattern of venture management (technological mission analysis) is described in a separate chapter. The hazards and advantages of each are identified and industrial examples provided of recent applications.

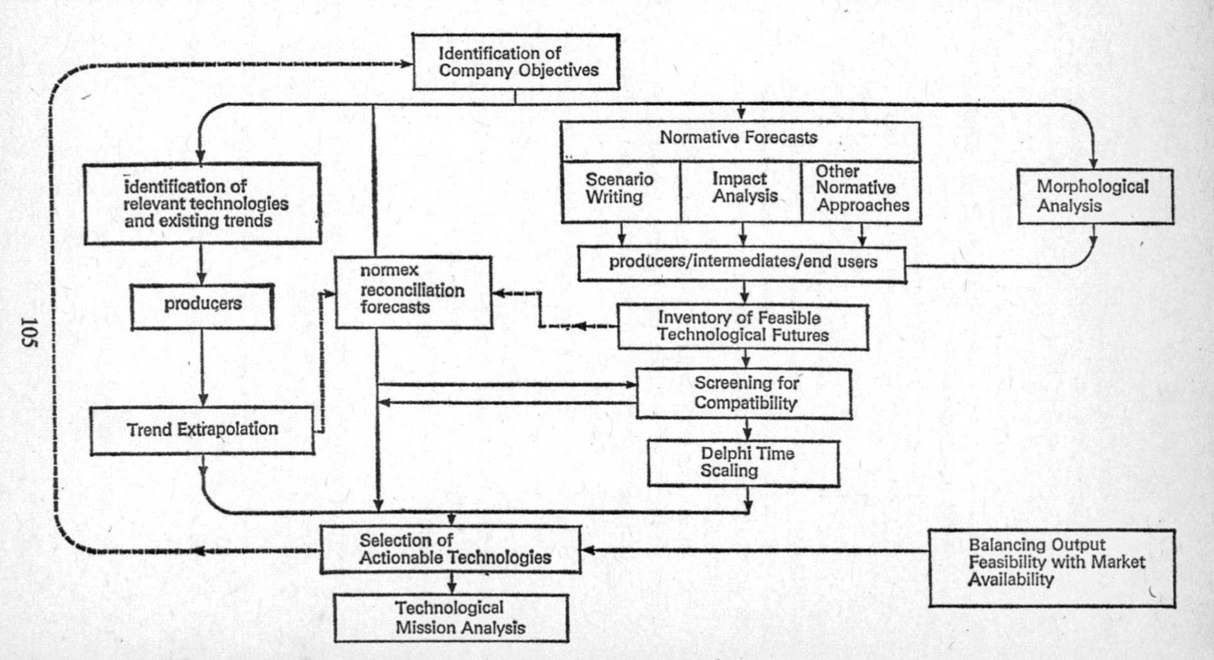

Exhibit 2. *A Sequence for the Deployment of Technological Forecasting in a Business*

CHAPTER 2.1

Extrapolative Approaches

EXTRAPOLATION is probably the only technique which has been widely employed in TF for some fifty years or more. It is also probably the only technique of which the layman coming to TF for the first time will have heard. The fact that it has been supplemented by a wide array of additional procedures is the outcome of its inability to cope with many aspects of future development. None the less it has substantial advantages and is widely employed in association with the planning of medium-term research and development and for the identification of major obstacles to forward movement in functional capability.

A word of caution is in order at this point for those who are familiar with the range of curve-fitting procedures which is normally brought to mind when extrapolation is mentioned. In the case of TF, the validity of detailed curve-fitting procedures is highly suspect. We are dealing with a broad sweep of development which can undermine some or many of the variables which have contributed to the historical trend we can observe. Direction is often more important, and more predictable, than the precise performance to be achieved or its exact time-scale.

Extrapolation of data in TF has made use of three major data sets:

(i) *functional capabilities*, independent of any particular technology, expressed either in direct performance terms or *figures of merit*, e.g., cost;
(ii) *capability of specific technologies* in accomplishing functional capabilities;
(iii) *scientific and technical findings*, as yet unrelated to either (i) or (ii).

We shall be particularly concerned in this chapter with (i) and (ii) and their interrelationships. Four main classes of trend curve

Figure 1. *Mechanization and Productivity in Coal*

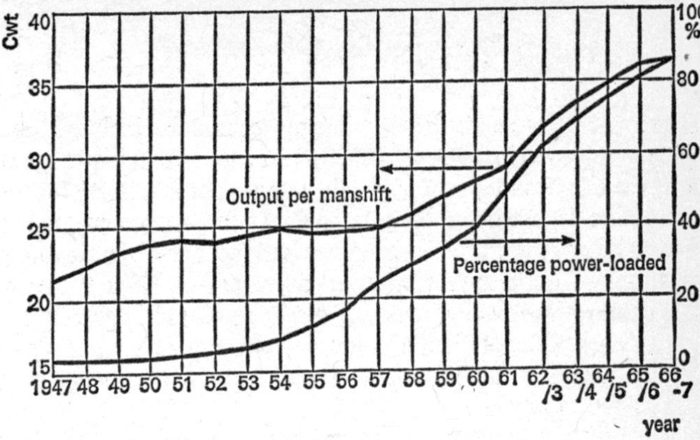

Cwt 40 35 30 25 20 15

100 % 80 60 40 20 0

Output per manshift

Percentage power-loaded

1947 48 49 50 51 52 53 54 55 56 57 58 59 60 61 62 /3 63 /4 64 /5 65 /6 66 -7

year

Source: *NCB Annual Report*, 1966–7

Figure 2(a). *Functional capability trend of energy-conversion efficiency for illumination (exponential)*

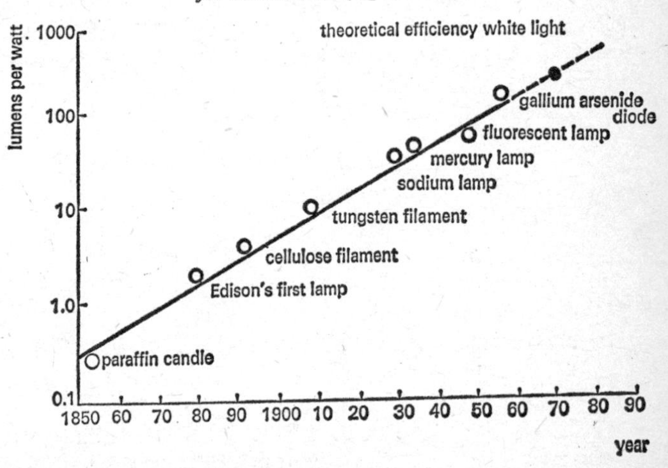

Figure 2(b). *Specific Technology* (*exponential constrained*)

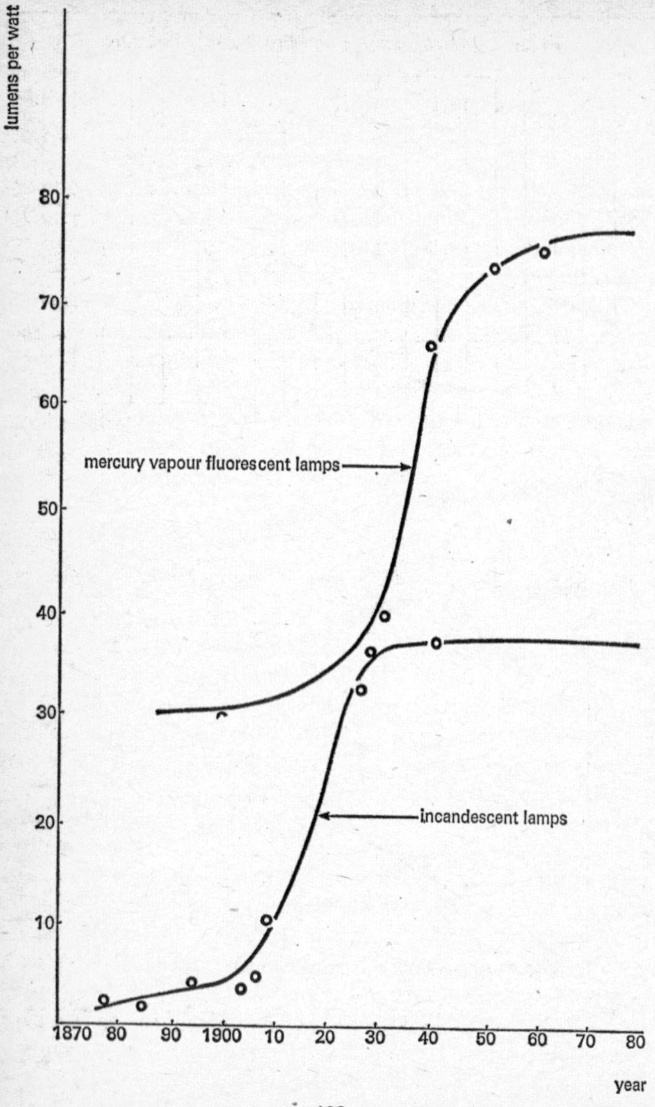

are relevant to our discussions and these are illustrated in Figures 1, 2(a), 2(b), 3 and 4.

Figure 1 describes a curve which we can term *linear increase with flattening*. On normal squared graph paper it describes a straight line. Several examples of such a trend have been instanced: the efficiency of thermal power plants or the mechanization of human work expressed in terms of the decrease in annual working hours per man. The relationship is also commonly seen in the growth in production of materials.

The most widely encountered curve, however, is that depicting *exponential growth, either with or without a constraint during the time period under review*. These two trends are instanced in Figure 2(a) and 2(b) respectively for functional capability in energy conversion efficiency for illumination (expressed as lumens per watt), and for two specific technologies contributing to that overall efficiency trend, i.e., incandescent and mercury-vapour fluorescent lamps.

It will be noted that two different plotting procedures have been employed: in Figure 2(a) semi-log paper has been used, which means that exponential growth, or a geometric rate of development, appears as a straight-line. If we had used ordinary squared graph paper, the result would be as demonstrated in Figure 2(b) – an S-shape. I have quite deliberately drawn them differently, because an S curve is a key phenomenon in TF which we shall be looking at further in a moment. A constrained exponential on semi-log paper will of course also give an S-shape, but it is much less pronounced, and barely looks like an S.

The third main class of curves is the *double exponential*, or even steeper increase, *with subsequent flattening*. This is characteristically encountered in the early phase of development in a field when R & D expenditures are concentrated very sharply indeed. This is the outcome of a research bandwagon effect, and where rapid growth in capability occurs it does much to emphasize the relative nature of time and resource allocation. Figure 3 shows the increase in output energies (in joules) of laser technology since 1962, when laser action in ruby was first demonstrated. Within four years there was an increase of five orders of magnitude.

Figure 3. *Single Pulse Energy in Lasers*
(multiple exponential growth with flattening)

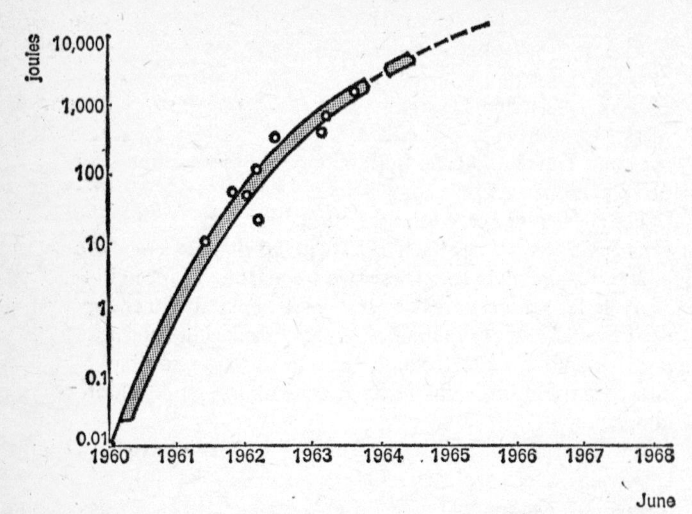

Source: Carter, J. R., *Lasers: How They Grew*, US Navy, IDP – 2109, 1967

The fourth and final main class of curves is the *slow exponential with sudden increase and eventual flattening*. This will be found especially in the extrapolation of functional capability, the sudden explosion occurring when a new specific technology hits the scene. Many exponential curves under the second or third classes would belong to this group if a longer time span was reviewed. Figure 4 demonstrates over a longer historical span man's speed trend. After century upon century of unmechanized, then mechanized, vehicles, powered mechanical vehicles have given a sudden exponential growth in speed capability.

One particular hazard of trend extrapolation is readily apparent here. The missile capability on the parameter selected, i.e., velocity/speed of light, is way ahead of trend. It is a multiple exponential growth situation. Of course, other unmanned missiles such as bullets have not been included but missile development can

Figure 4. *Speed Trend of Man*
(slow exponential with sudden increase and eventual flattening)

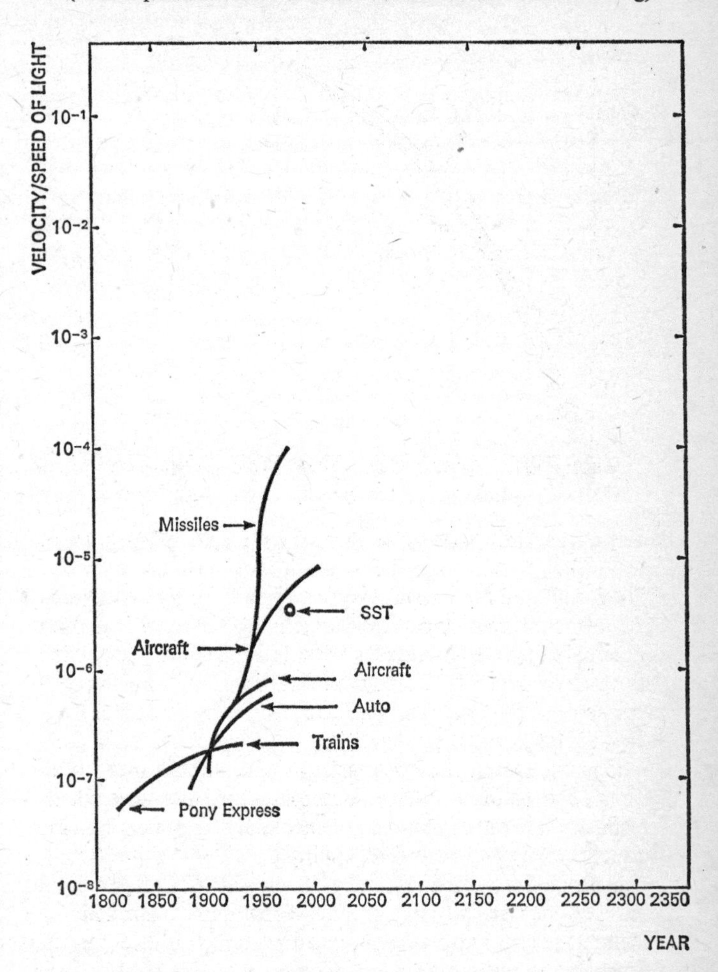

Source: Samaras, D. G., cited in Bright, J. R. (ed.), (1968), p. 104.

be deemed to have upset the balance of the extrapolation in a manner similar to that found for laser technology. Massive R & D resources were committed to it in a way quite untypical of the emerging historical pattern.

CONSTRAINTS AND ENVELOPES

We have deliberately examined the main classes of curves which have been identified and their data inputs, before discussing 'how the process of extrapolative TF goes along'. We can now more meaningfully look at the rules, stated most clearly by Ayres and Lenz in J. R. Bright's edited papers, *Technological Forecasting for Industry and Government.*

The fundamental problem in extrapolation is twofold:

(i) what criteria should we employ in deciding what to extrapolate?
(ii) when can we happily (or naïvely) extrapolate history, and when must we posit the possibility of points of inflection, of changes in the rate of growth of capabilities?

The first question is easily answered in principle but is more intractable in practice. Any salient variable may be taken provided it can be defined independently of any specific class of devices. Specific thrust is such a macro-variable, whereas combustion-chamber temperature is not. This latter is a micro-variable which presupposes some particular technological approach to delivery of a capability. An analysis of erroneous forecasts by earlier futurists shows specificity of device as perhaps the single most vulnerable element in their predictions. Aggregation of devices into broad functional classes is therefore the guideline for parameter definition. Specific devices can be, and of course are, considered separately such as the system elements of an aircraft – its engine, guidance, airframe and so forth. But they will only facilitate meaningful forecasts if they are to act as upper constraints on functional capability. This is hazardous country, however. The development of the jet engine transformed any analysis of engine performance and constraints under prior technology. Lasers provide an equally important reminder of constrained thinking with an unconstrained capability. The optical theorem suggests that the image cannot be brighter than the source intensity of light. Rigorously stated, of course, this is correct, but amplification by stimulated emission, i.e., the laser, upset many forecasts on the rigorous basis.

The importance of aggregative parameter identification is demonstrated in Figure 5, which was prepared at ICL. If para-

Figure 5. *Enveloping of Capability of Specific Devices to give Overall Functional Capability Trend*

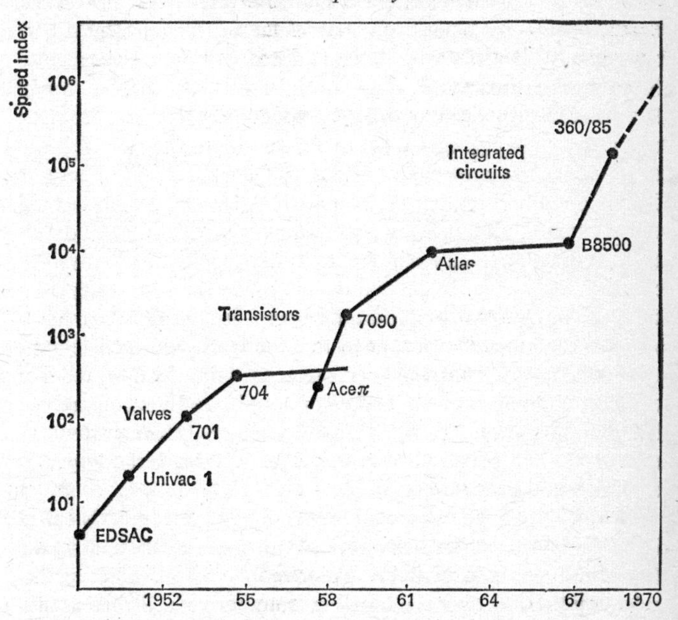

Source: Hall, P. D., in Wills, G. S. C. et al., (1969), p. 191.

meters for each machine in the sequence, as for each specific technology, were developed, they would have forecast a flattening growth in capability. The enveloping trend – *the envelope curve* – does not behave that way.

Replete with a suitable macro-variable, the extrapolator must now *determine whether it is extensive or intensive*. This implies identifying whether the macro-variable, or parameter as it is frequently termed, is approaching an intrinsic or physical constraint. If it is, then it is termed intensive, and if not, extensive.

Absolute (physical) constraints are easiest to describe – Einstein speed (the velocity of light) or absolute zero temperature. Intrinsic constraints are either the 100 per cent or unity boundary in ratios such as input–output, or constraints imposed by the human body, e.g., size or the environment, or escape velocities from gravitational fields. An extensive variable can well be constrained by non-technological factors. Aircraft speed will be a compromise between size, turnround time, noise limitations and so forth. Extensive macro-variables will also frequently be time-constrained; that is to say no aggregated capability development can take place until a tangential area has advanced. We shall see later, under our discussion of normative techniques for relevance analysis and under technological mission analysis, how vital the phased development of component technologies is for efficient TF. Analysis of constraints frequently points the way for rigorous tangential development work. Resolution in cameras is an extensive macro-variable constrained over time by film speed and lens focal length; laser coherence is a parameter constrained in time by the quality of crystals; computer central processor speeds are time-constrained by miniaturization technology for computers. Hovercraft technology is time constrained by skirt and trunk technology, as demonstrated in Figure 6.

The air-cushioned vehicle (or hovercraft as it is popularly called in the UK) will substantially improve its performance against the factor of merit, as skirt and trunk technology affords further considerable benefits over the bare-bottom approach. Problems still remain in finding the most suitable materials for the skirts, however, but that these will be overcome can be confidently predicted.

There is a third class of constraints which is perhaps a special class of the extensive constraints already described. It is the time-dependent limitation on performance imposed by the operation of the market place. Costs of production are partially determined by

Figure 6. *Hovercraft Performance with Skirt and Trunk Technology as a Time-dependent Constraint*

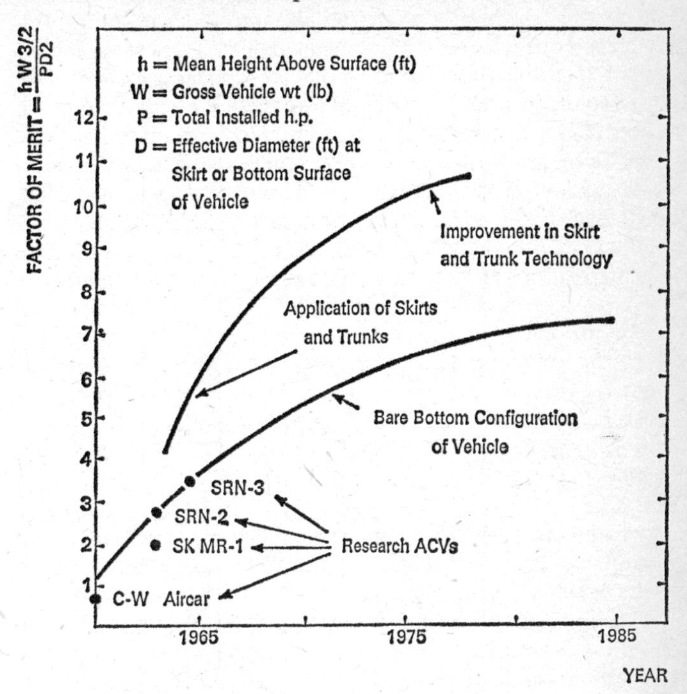

Source: Naval Ship Research and Development Centre

the level of functional capability, other things being equal, but production cost is also determined by the level of demand. At too high a price there will be no demand; as price falls more will generally materialize. Figure 7 indicates the price–quantity relationship for thermoplastics in the USA, UK and EEC.

Ayres cites the case of long-distance operating voltages for electrical transmission lines. This is not determined by the optimum capabilities of conductors and insulators but by a variety of time-dependent market place constraints – development costs,

Figure 7. *Price/Quantity Relationship*

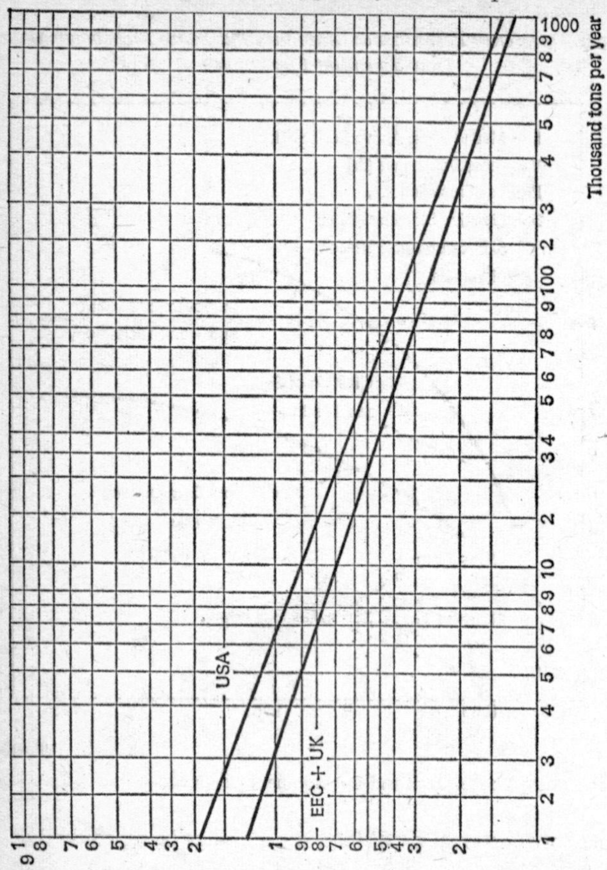

EEC + UK Slope – 0.345

USA slope – 0.40

In each case PTFE and nylons are much greater than curve indications

13 items listed

1. Polyacetal; 2. Polycarbonate; 3. Cellulosics; 4. High pressure polyethylene; 5. L.p. polyethylene; 6. Fluorinated polymers; 7. Nylons; 8. Polypropylene; 9. Polystyrene; 10. ABS-SAN copolymors; 11. PVC; 12. PVA etc.; 13. Acrylics

Source: Gregory, S. A., 'Trend Extrapolation', in *Design and Innovation Group Symposium*, (1969)

capital costs and amortization of land purchased for right-of-way and safety, and anticipated operating costs and savings based on power, length of lines, number of taps, average load factor, etc. Until such market-place constraints are removed, the incentive for any advance in capability is dulled, and extrapolation must beware of overlooking this phenomenon.

Figure 8. *Falling Cost Forecast for Desalination Technology*

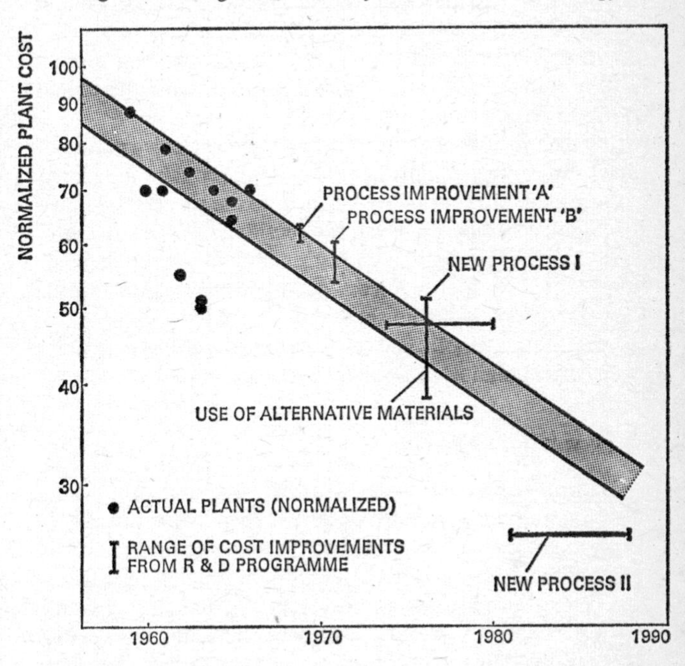

Source: Garrett, T., in Wills, G. S. C., et al., (1969), p. 235.

There will be certain cases where the identification of real time is less precise and the next step of progress will be dependent on one or more parameters. Such a parameter-dependent trend is often felt to be the case in the technology for desalination of water. The cost of power and the diffusion of membranes are two which are obviously of vital significance. Their joint occurrence will determine the performance of constructable diffusion desalination plants. The Ministry of Technology's Programmes Analysis Unit was bold enough to attempt the forecast of cost to 1990 which appears as Figure 8. Plant costs have been normalized to 2 million gallon/day units with a performance ratio of 8:1.

Such forecasts could easily be upset if the expected new process failed to live up to expectations, and it is this which hitherto had led others to refrain from such forecasts.

Having identified the relevant intensive and extensive nature of our macro-variable, and duly noted any of the constraining influences during the time period with which we are preoccupied, two final rules must be obeyed. First, the sensitivity of the parameter must be examined; second, we must undertake similar steps for any likely interacting effects from outside technologies. Electric fuel cell technology is obviously such an interacting influence on many traditional technological fields. In the light of these few steps a sensible extrapolation can usually be made.

TECHNOLOGY PRECURSORS AND MULTIPLE TRENDS

There is a great deal of evidence to indicate that in the short term the search for a precursor event or development is worthwhile. In the assessment of levels of investment in manufacturing industry and the future demand for machine tools, for instance, changes in business confidence are significant precursors. Ideally, we would search for a perfectly parallel lagged relationship. Only a few instances are cited, the most commonly seen has been developed by Lenz. It shows how spin-off from military aircraft development moves into the civil sector. Whether this precursor approach will continue to apply in these sectors is open to question; certainly the time lag is widening, from nine years in 1930 to thirteen years in

1990. At any point in time transport speed is approximately 52 per cent of combat aircraft speed. This is indicated in Figure 9.

Figure 9. *Speed Trends of Combat Aircraft* Versus *Speed Trends of Transport Aircraft, Showing Lead Trend Effect*

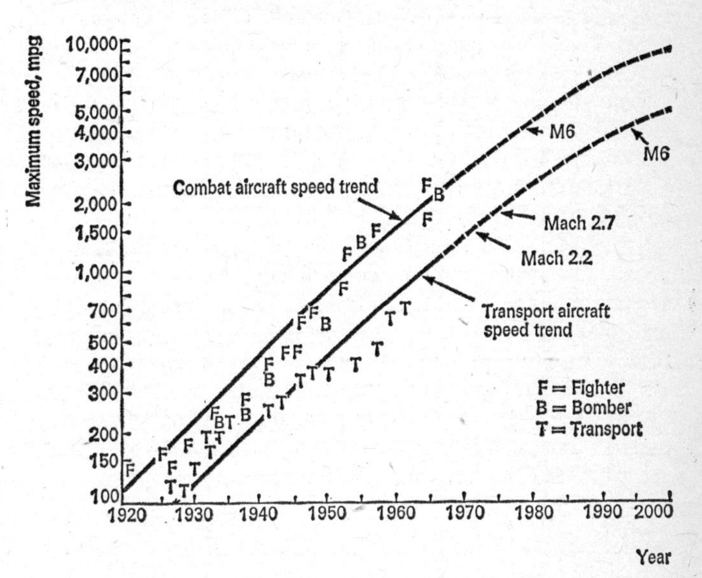

Source: Lenz, R. C. Jnr., (1962), in Bright, J. R. (ed.), (1968)

Mumford has undertaken a precursor-style analysis for aerosol units produced in the UK which is given in Figure 10.

Lenz goes on to demonstrate a further valuable device in TF, by the manipulation of algebraic relationships between parameters. He observes that certain variables can be independently extrapolated from good historical data, e.g., domestic trunk airline passenger miles, plane miles, seating capacity and load factor, but

Figure 10. *Aerosol Units Produced; Total and* Per Capita

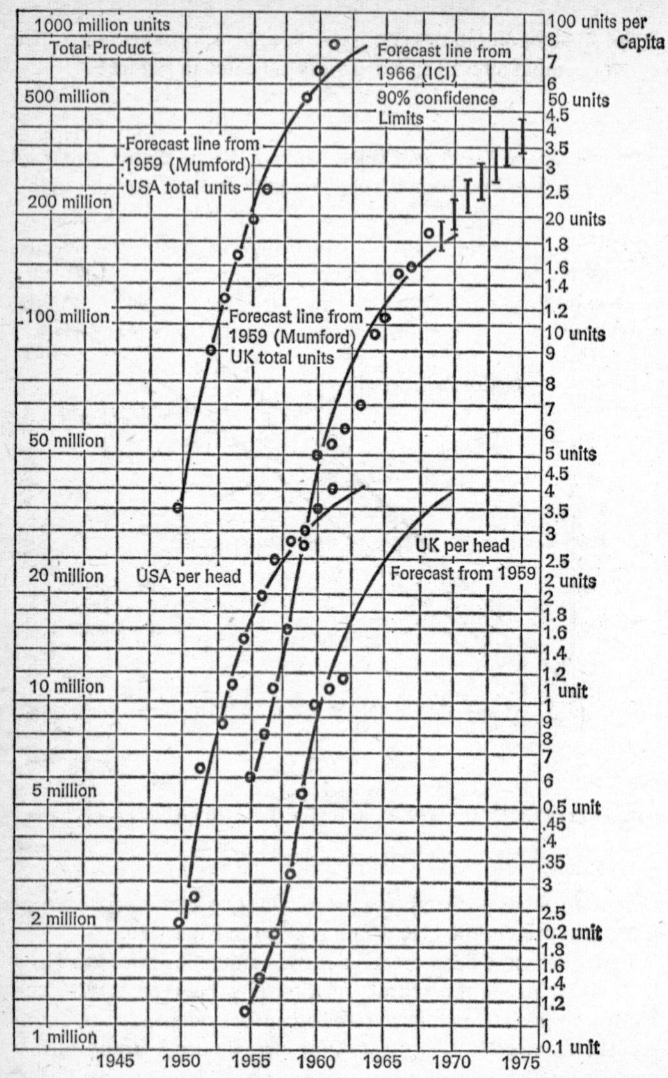

Source: Mumford, L. S., 1963; cited by Gregory, S. A., in *Design and Innovation Group Symposium* (1969)

that they are *not* independent of one another. In fact, they are related as follows:

$$\text{Plane Miles} = \frac{\text{Passenger Miles}}{\text{Seating Capacity} \times \text{Load factor}}$$

The perhaps somewhat surprising conclusion emerges from such analysis that plane miles will actually fall after 1970 if a load factor of about 60 per cent is assumed. The important point to note, of course, is the fixed relationship of these various trends so that the extrapolation of any one variable must consider the effects of that on the others. This is demonstrated in Figure 11.

Figure 11. *Domestic Trunk Airlines; Multiple Trend Forecast*

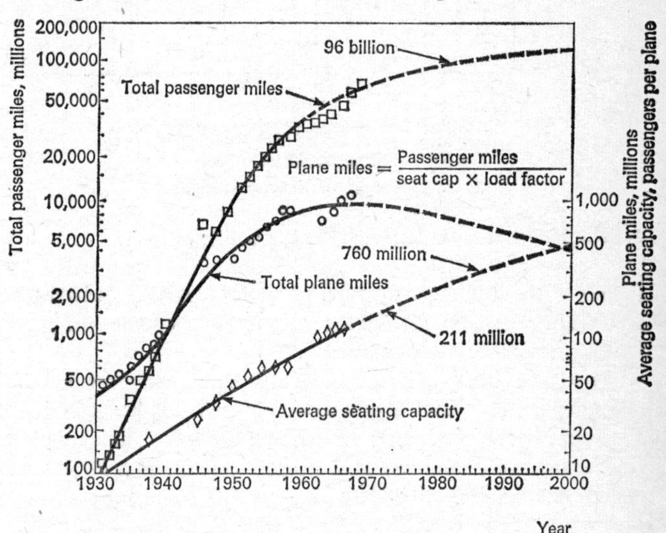

Source: Lenz, R. C. jnr., (1962), in Bright, J. R. (ed.), (1968)

FLOYD'S PHENOMENOLOGICAL MODEL

Dr Floyd has made the most substantial attempt to date to develop a mathematical model which permits trend extrapolation of

Figure 12. *Technique Availability in Relation to Level of Figure of Merit*

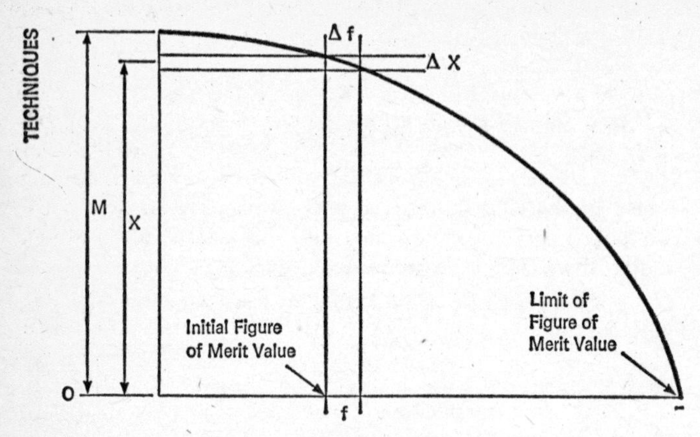

Source: Floyd, A. L., in Bright, J. R. (ed.), (1968), p. 96

figures of merit in given areas of technology. It also identifies the probabilities associated with the forecast and estimates the effects of competitive technologies.

Floyd's basic approach involves calculating the probability that a figure of merit can be improved through applied effort. He assumes that there are X techniques available that could be potentially successful and M is the universe of all possible techniques that could be considered.*

Then: $$P(f, 1) = \frac{X}{M}$$

that is to say, the *a priori* probability of success per attempt. If

*This discussion is hopefully within the comprehension of the non-mathematically minded reader. I have endeavoured to make it meaningful in both symbols and prose. May I exhort the symbol-rejectionists to read on?

there are W research workers, performing W effort per unit time to improve the figure of merit, the probability of exceeding a given level in the figure of merit in a given time span (Δt) is:

$$P(f, \Delta t) = 1 - \left(1 - \frac{X}{M}\right) NW (\Delta t)$$

i.e., the probability is 1—the probability of not making the improvement.

The value of $\frac{X}{M}$ can be estimated in terms of figures of merit values. Any available technique can improve the figure of merit beyond its initial value. As improvements are made the available store of techniques gradually decreases. The curve can and does take many forms, but is normally similar to the curves for absorption phenomena. As a first assumption,

Floyd suggests that the number of techniques remaining available to improve a figure of merit is proportional to the number already absorbed to achieve the extant figure of merit, or:

$$\frac{\Delta X}{\Delta f} = -K(M - X)$$

where Δf is the change in the figure of merit and K is a constant.

Integrating this equation between f and the limiting value F, in Figure 12, and X = X and X = 0 yields:

$$\frac{X}{M} = 1 - \exp[-K(F - f)]$$

Substituting in the earlier expression for $P(f, \Delta t)$, we obtain

$$P(f, \Delta t) = 1 - \exp[-(F - f) KNW (\Delta t)]$$

which is the equivalent of saying that the probability for achieving f for any total time t is given by the probabilities of not achieving f and subtracting this value from 1 $[1 - P(t)]$.

$$P(f, t) = 1 - \exp.[-(F - f) \int_{-a}^{t} KNWdt]$$

In this last equation, the sum of all KNW (Δt) elements is replaced by the integral. This would allow probability calculations if the integral could be evaluated but generally Floyd concedes it cannot.

He proceeds accordingly to examine the variables within the integral. K is expected to vary over time but should not do so too widely since it is dependent on the general growth of technology. N should be a well-behaved function of time. However, W can be varied at will and widely particularly in crash programmes and where transfer effects occur from other fields. He suggests taking account of this as follows in a first order approximation:

$$W = W_0(t)\,(f - f_C)^P$$

where $W_0(t)$ represents the constantly growing number of workers available and f_C is the figure of merit of some competitive technology. The greater the difference between f and f_C the more workers will enter the field. This then yields a final expression:

$$P(f, t) = 1 - \exp\left[-(F - f)\int_{-a}^{t} (f - f_C)^P\, T(t)dt \right]$$

where T (t) is a composite slowly varying function of time.

$T(t) = K(t)\,N(t)\,W_0(t)$

Floyd has demonstrated that the time behaviour of the figure of merit F(t) can be determined only in terms of the function:

$$\int_{-a}^{t} \frac{T(t^1)dt^1}{\ln 2} = g(t).$$

He uses the device of specifying a 50 per cent probability of success. The variables can be separated to yield:

$$P(f, t) = 0{\cdot}5 = 1 - \exp\left[\frac{-0{\cdot}6931\,(C_1t + C_2)}{Y + \ln(Y - 1) + C_2}\right]$$

$$\text{where } Y = \frac{1 - fc/F}{1 - f/F}$$

C_1, C_2 = constants

t = time

and the integral T(t) has been replaced by a constant times the change in time for finite ranges of time considered.

Three values are now required to use this equation for extrapolation: F, the limiting figure of value must be calculated; fc must be estimated; C_1 value as a constant must be determined from at least two data points.

By taking the equation and setting P(f, t) equal to a constant, and then taking the logarithm of each side:

$$Y + \ln (Y - 1)^{tc}2 = -(C_1 . t + C_2)$$

From this revised form, a universal nomograph can be constructed as shown in Figure 13. It shows a plot of

$$Y + \ln (Y - 1) \text{ or } C_1 . t$$

versus

f/F

When fc/F = zero, this represents the general technology trend curve, i.e., when the competitive technology figure of merit value is 0. The value of fc/F acts as a scaling factor for f/F and its value has been plotted across the top of the nomograph with a value of 1 as reference.

Figure 13. *Floyd's Nomograph for calculating C_1.t and thence to Trend Extrapolation*

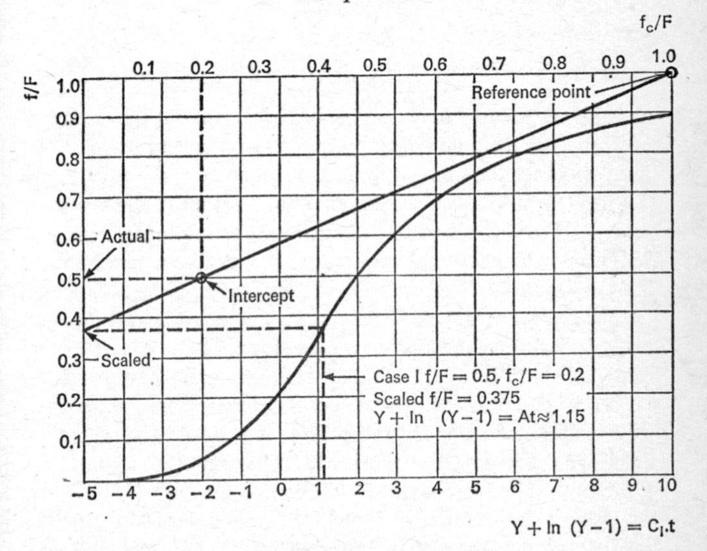

Source: Floyd, A. L., in Bright, J. R. (ed.), (1968), p. 99.

The nomograph is used as follows:

(i) Determine value of fc/F and f/F and plot the intercept.
(ii) Draw straight line from reference point (fc/F = 1·0) through intercept to the f/F axis. This represents the scaled value of f/F.
(iii) Read value of C_1.t at point where scaled value of f/F is projected on to the curve.

In Figure 13, this sequence of steps is indicated for:

f/F = 0·5
fc/F = 0·2
scaled f/F = 0·375
C_1.t = 1·15

Any trend can be thus extrapolated and Floyd reports good fits to several data sets; the trend fitted in Figure 4 (p. 111) was in fact made by Floyd's phenomenological model.

LEARNING CURVES AND BUDGETARY ALLOCATIONS

The value of extrapolative approaches is to map out the nature of likely territory through which businesses will have to travel in the future, whether they like it or not. With such knowledge, more informed decisions can be taken about resource dispositions, particularly to various specific contributing technologies within an overall field of functional capability. The problem can be illustrated in Figure 14. At what point in time are resources reallocated to technology B from A? Some have suggested that allocation should be made in proportion to the slopes of the curves at any particular point in time, but this is certainly myopic, particularly at time t2. The probabilities of successful development are obviously in quite the opposite direction. An absorption approach might be more logical. Budgets for R & D will need to be allocated on the basis of *expected* rates of advance, not on an historically recorded section of a trend. Switchover points must none the less be identified, and there is an ever present danger of technology B following route B_L rather than B_H.

Figure 14. *Opportunities and pitfalls of technology switchover forecasts*

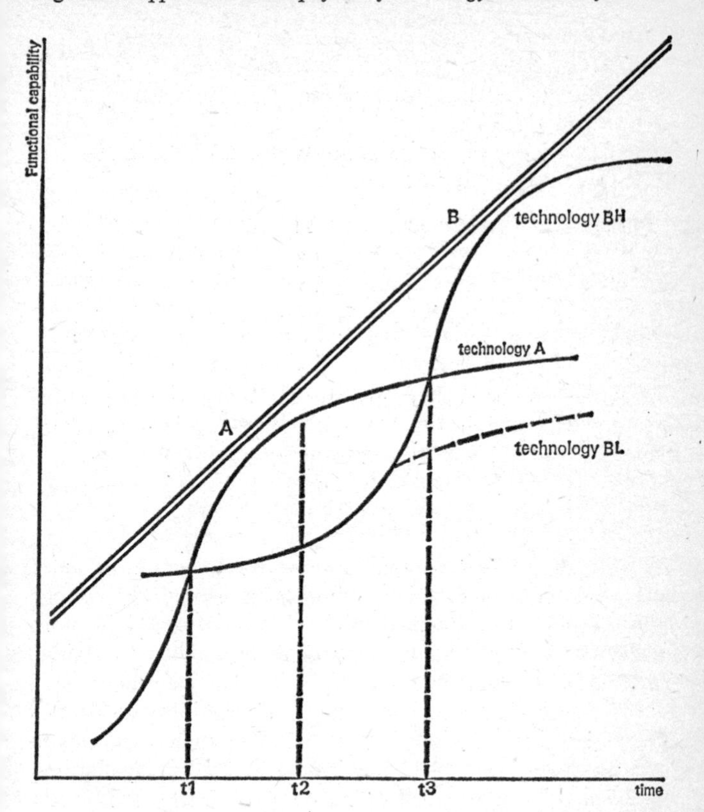

In practice, as we have indicated at an extreme with laser technology, R & D expenditure may well be very high at the onset of a new technology and its rate of initial increase will be greater than that for the functional-capability achievement. This will often lead to a cut-back in R & D projects just as an exponential growth is materializing. A sensible understanding of growth phenomena can do much to avoid such futile oscillations within a total R & D system but will not, of course, completely eliminate the familiar patterns inherent in industrial dynamics.

Technological forecasters are in the long term expected to be subject to excessive caution. We have already noted that science-fiction writers are generally better forecasters than scientists or technologists. In extrapolation, we can correct for this. *Learning curves* are identifiable as a technique for accomplishing this. Some detailed effort has been put into an examination of development costs, and time, and estimated production costs at various moments over the life of a development. The British public has become accustomed to this process of learning in relation to aircraft estimates over the past two decades or more; most notoriously perhaps with Concorde, whose cost estimates have now nearly quintupled from their 1962 start at about £165 million. The general learning relationship for production costs has appeared in many cases to be linear as development time elapses rather than manifesting a hoped-for early learning process. Quite simply, until the technical elements of a system are known, advanced technology cost-estimating has proved to be almost impossible at a project outset. This is a salutary observation for the first-to-market technology firms in business today and it is reflected in the very high rates of return expected to compensate for such risks.

One final and obvious comment on the monitoring and control of parameter extrapolation is that it should be updated. The variance between forecast trends and actual capabilities or figures of merit will quite frequently constitute a most powerful input for the next round of extrapolation. The analysis of such variance may often provide the key to substantially more accurate extrapolation subsequently.

APPENDIX

The Estimation of Confidence Intervals in an Extrapolative Forecast

IN making exploratory forecasts, it is desirable to evaluate the uncertainty associated with the forecast. This can be accomplished by the following procedures.

First, a regression analysis is carried out which establishes an

Figure 2.1A. *Example of the use of Confidence Intervals in Extrapolative Forecasts*

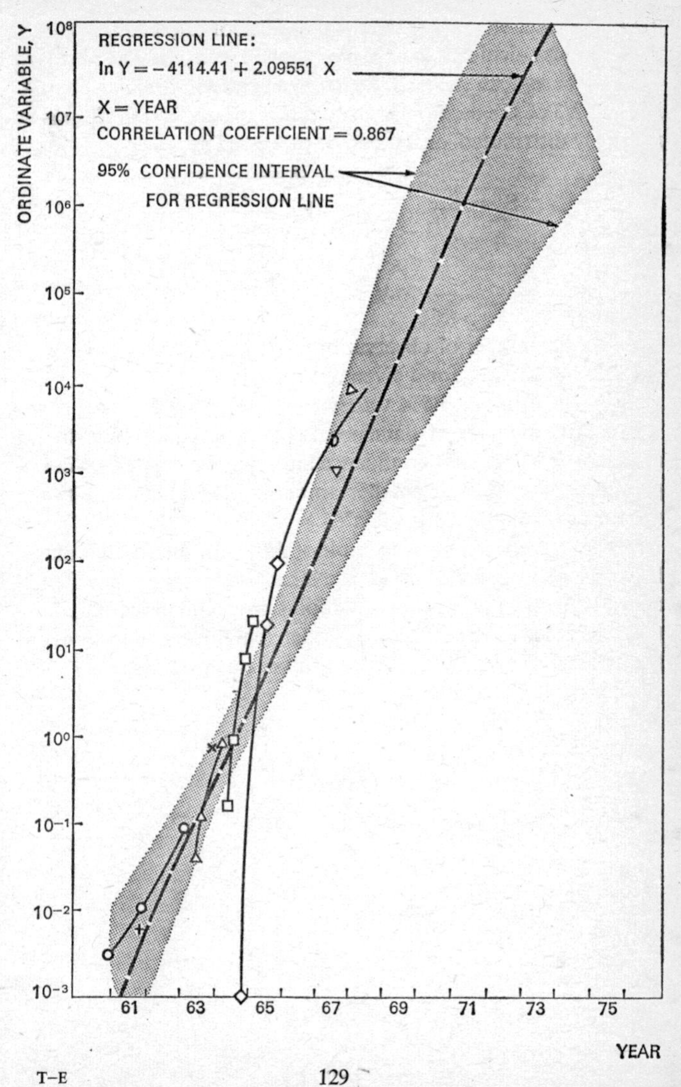

equation for the trend line through the data from which the exploratory forecast is made. The standard error of the estimate of Y on X, S_{yx}, is calculated from the regression analysis. The standard error of the regression line S_y defines the band within which approximately 2/3 of the data points would be expected to fall if they are normally distributed, and it can be expressed as

$$S_y = S_{yx} \sqrt{\frac{1}{n} + \frac{(X_i - \bar{X})^2}{\sum_{i=1}^{n} (X_i - \bar{X})^2}}$$

where

n = the number of observations
X_1 = the X value for a given data point
$\bar{X}$ = the mean of the X values

If a large number of data points (greater than about 30) are available, a 95 per cent confidence limit may be defined by multiplying S_y by 1·96. For smaller amounts of data, the multiplier should be taken from a 't table'.

Figure 2.1A presents an example of the use of this methodology. It can be seen that the confidence interval is most narrow (confidence is highest) around the midpoint of the historical data and the confidence interval becomes wider (confidence decreases) as the data are projected farther into future time periods.

CHAPTER 2.2

Morphological Analysis

MORPHOLOGICAL analysis is one of two major techniques of TF for which we are indebted for inspiration to classical Greece. The other is the Delphi method discussed in Chapter 2.4 which employs a technique which is analogous to the Oracle at Delphi. Plato and Aristotle are the two names associated with the origin of analytical approaches to what they termed *form*. (The Greek word for 'form' is *morphé*.) Gregory, in particular, has been assiduous in tracking down the early emergence of this approach and I have relied on him for much of the historical detail I give here.

Aristotle, in his *Politics*, demonstrates the morphological approach as follows:

> If we were going to speak of the different species of animals, we should first of all determine the organs that are indispensable to every animal; as, for example, are the organs of sense and the instruments of receiving and digesting food, such as the mouth and stomach, besides organs of locomotion. Assuming that there are only so many kinds of organs, but that there may be differences in them – I mean different kinds of mouth, and stomach, and perception and locomotive organs – the possible combinations of these differences will necessarily furnish many varieties of animal. For animals cannot be the same which have different kinds of mouth or ear. And when all the combinations are exhausted there will be as many sorts or animal as there are combinations of the necessary organs. The same, then, is true of the forms of government . . .

Medieval philosophers and theologians made further use of the method. Ramon Lull (cited in Martin Gardner, *Logic Machines and Diagrams*) postulated nine attributes of God, which infused the whole of creation. He sought to develop an art for the conversion of Jews and Muslims, by combining each of the attributes with various levels of nature, and employed geometrical devices. Figure 1 demonstrates the Attribute Diagram and levels of nature which Lull developed, along with binary combinations.

Figure 1. *Lull's Attribute Diagrams with Binary Combinations for Sixteen Terms*

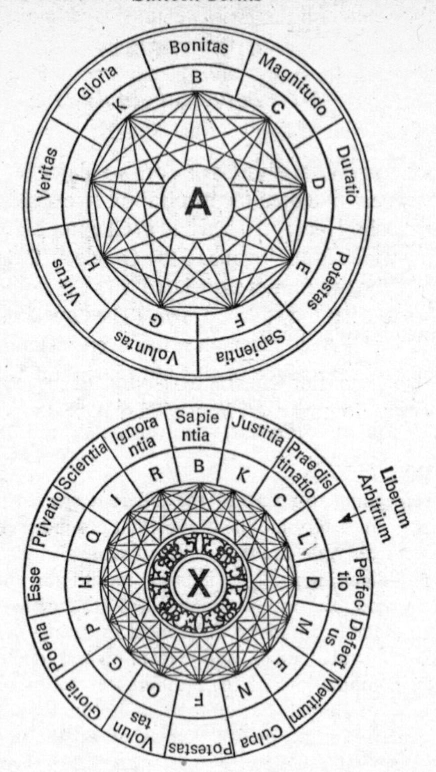

BC	BD	BE	BF	BG	BH	BI	BK	BL	BM	BN	BO	BP	BQ	BR
	CD	CE	CF	CG	CH	CI	CK	CL	CM	CN	CO	CP	CQ	CR
		DE	DF	DG	DH	DI	DK	DL	DM	DN	DO	DP	DQ	DR
			EF	EG	EH	EI	EK	EL	EM	EN	EO	EP	EQ	ER
				FG	FH	FI	FK	FL	FM	FN	FO	FP	FQ	FR
					GH	GI	GK	GL	GM	GN	GO	GP	GQ	GR
						HI	HK	HL	HM	HN	HO	HP	HQ	HR
							IK	IL	IM	IN	IO	IP	IQ	IR
								KL	KM	KN	KO	KP	KQ	KR
									LM	LN	LO	LP	LQ	LR
										MN	MO	MP	MQ	MR
											NO	NP	NQ	NR
												OP	OQ	OR
													PQ	PR
														QR

Goethe picked up the Aristotelian theme again when, in 1790, he advanced the notion that members within a large group present something in the nature of variants in a common plan. He employed the term 'morphology' to describe the study of the structure of living things, their relationship with other living things, how their structural relationships arise, and the factors that go into their production. Darwinism and the emergence of taxonomic approaches were a natural outgrowth. Darwin's basic propositions in *The Origin of the Species*, published in the middle of the nineteenth century, can be cited here, in the context of technological innovation and its successful commercialization, with considerable relish:

(i) that gradations in the perfection of any organ or instinct either do now exist or could have existed, each good of its kind;
(ii) that all organs and instincts are, in ever so slight degree, variable;
(iii) there is a struggle for existence leading to the preservation of each profitable deviation.

Franz Reuleaux is credited with the first direct application of morphological analysis to technological forecasting and speculation. He published his ideas in *The Kinematics of Machinery* toward the end of the nineteenth century whilst working in Berlin as Director of the Royal Industrial Academy. He defined a machine as 'a combination of resistant bodies so arranged that by their means the mechanical forces of nature can be compelled to do work accompanied by certain determinate motions'. These determinate motions are occasioned by kinematic chains, and a closed kinematic chain, which has one link made stationary, is a mechanism.

He discovered that all the machines he had examined fell into two great classes: they were either place or form-changing. Each machine, in addition to its driver and workpiece, had the following mechanisms:

(i) a main train in which receptors and tools must exist;

(ii) director, with sub-division for supply and discharge;
(iii) regulator, with sub-division for stop gear; and
(iv) gearing, or transmitting gear.

Reuleaux commented, of his thesis, that 'we have developed and systematized a method not infrequently made use of by practical men, and given it distinct form, [rather] than introduced a completely new idea'. From such a structural analysis he proceeded to direct or indirect synthesis, which involved the detailed exploration of possible alternative routes for machine construction. Figure 2 illustrates this activity.

Figure 2. *Reuleaux Diagram of Kinematic Synthesis*

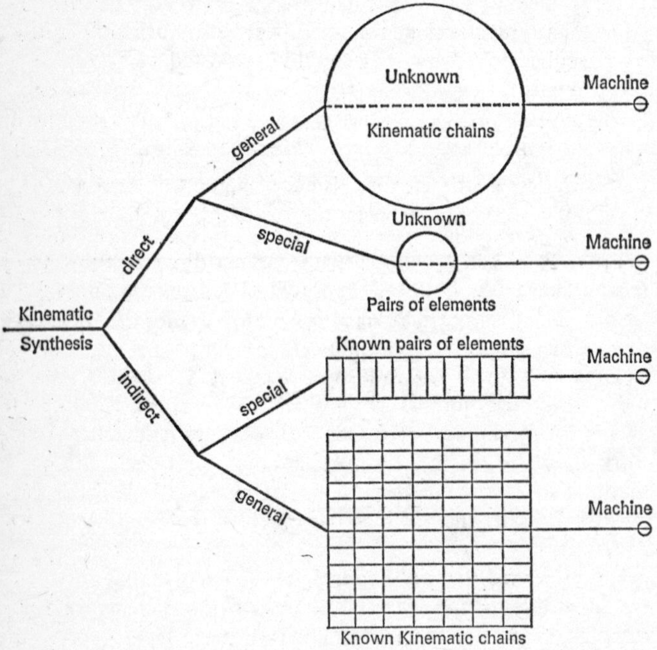

It was with such illustrious antecedents that Fritz Zwickey, the famous astrophysicist, formulated and popularized morphological analysis as a technique in 1948. His formalized method was intended 'to identify, index, count and parameterize the collection of all possible devices to achieve a specified functional capability'. It can be deployed for identifying and counting all possible means to an end at any level of abstraction or aggregation. Zwickey expounded the following rules:

(i) the problem to be solved, or the functional capability to be achieved, must be stated with great precision;
(ii) the characteristic parameters must be identified (this is to a considerable degree an automatic consequence of the precise statement at (i). Completeness, however, is the problem and only the analyst's capabilities can ensure it);
(iii) each characteristic parameter must be subdivided into distinguishable cases or states (say P_n^1, P_n^2, P_n^3. . .); or more frequently a continuum of values which must be meaningfully clarified into ranges or regimes, e.g., sub- and supersonic speeds may be seen as clear cases whereas supersonic and hypersonic regimes are not clearly divisible;
(iv) some 'universal' method of analysing the performance and feasibility of the various combinations is required, although Zwickey concedes that this is no easy task. Indeed it is seldom practicable, and a variety of extensions to his original formulation have attempted to grapple with the data deluge which normally arises.

Zwickey was a pioneer in the field of jet-engine development, and it was in this area that he made his original analysis. It is shown as Figure 3.

It focuses on the totality of all jet engines operating in a pure medium containing simple elements only and being activated by chemical energy – thereby fulfilling rule (i). Eleven characteristic parameters were identified, and Figure 3 indicates the distinguishable states suggested. The analysis yields 36,864 distinguishable combinations (i.e., $2 \times 2 \times 3 \times 2 \times 2 \times 4 \times 4 \times 4 \times 3 \times 2 \times 2$). One interesting combination from that vast array is plotted, a ramjet deriving its energy entirely from the surrounding medium. Zwickey went on to indicate that only 25,344 combinations were

Figure 3. *Morphological Analysis of a Jet Engine*

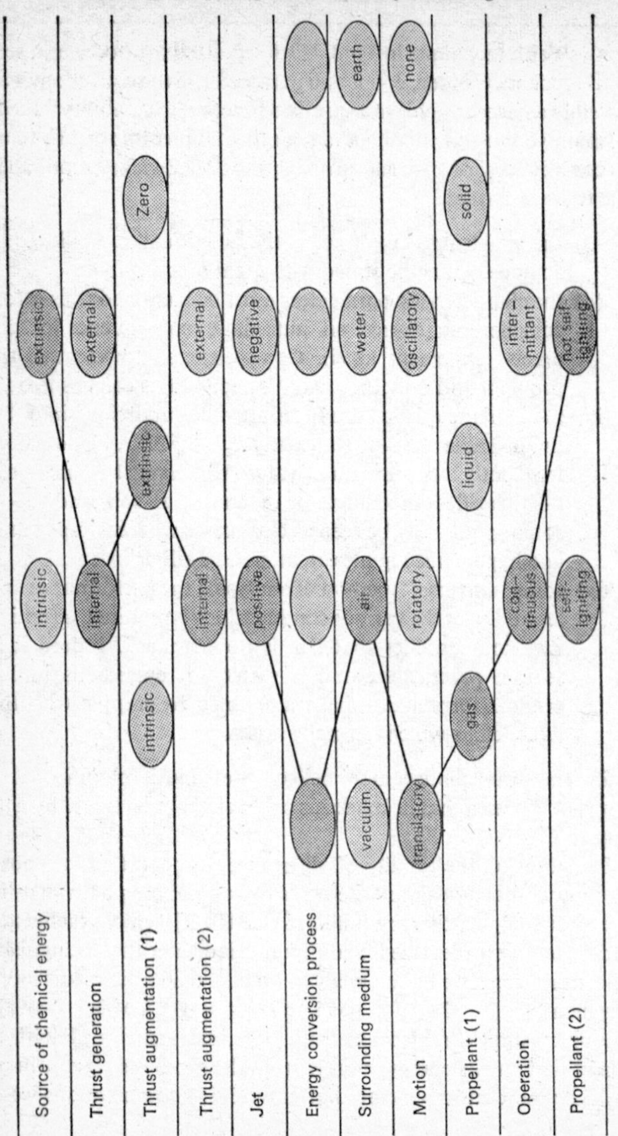

Source: E. Jantsch, 'Forecasting The Future', *Science Journal*, October, 1967, and 'Morphology of Propulsive Power' by Fritz Zwickey.

meaningful, however, since some were self-contradictory, e.g., any distinction between internal and external thrust is invalid where thrust augmentation is zero, and extrinsic chemically active mass cannot apply in a vacuum as the surrounding medium. (The energy conversion process cases are postulated as adiabatic, isothermal, etc.)

Ayres has pointed out that at least one particularly important characteristic parameter was omitted thus emphasizing the problems of adhering to rule (ii). He emphasizes that an important distinction can be made between whether combustion occurs at subsonic, supersonic or hypersonic speeds. Such an additional parameter with three cases will probably treble the combinations potentially displayable, and the problems under rule (iv) will be increasingly apparent to the reader.

The difficulties do not end even there, however. As Ayres goes on to point out, as demonstrated in Figure 4, the variety of possible ways of propelling a vehicle, even when limited to a chemical-fuelled jet, is surprisingly large. Figure 4 is an elaborate check-list for the organization of a broad study and as such is very valuable in business as a disciplined approach. It should be noted that some lines do not connect to each and every row of boxes. There is, for instance, no known way of charging a battery or of creating a jet with a tightly wound spring. Some feasible but economically non-viable connections have also been omitted such as the employment of nuclear combustion to support a photovoltaic cell. Its efficiency as a convertor would be ludicrously small, yet it has a similar basis to the solar cell mechanism.

Morphological analysis has not been solely applied in the field of aviation or defence research contract work, however, although the great majority of published studies are of a military nature. Norris has compiled perhaps the most extensive inventory of published studies which demonstrates the diversity of application since 1958 – e.g., ocean transport, cigarette lighters, car automatic transmission, excavators, lightweight wheels, domestic central heating, barge tanks, land storage tanks and so forth. I myself have been involved with applications for textile product ranges, agricultural fertilizers, functional fluids, retail sales data capture,

Figure 4. *Sources of propulsive work after Ayres. (Figures in parentheses below some items refer to the number of possibilities.)*

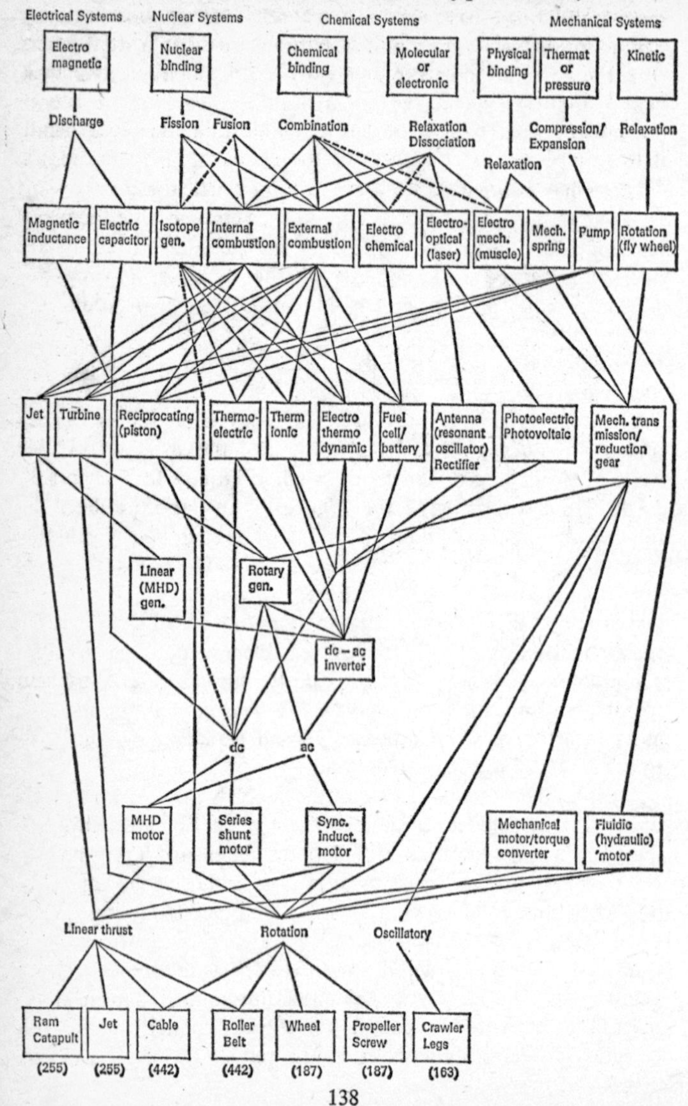

marine communications, rear-projection screening and, more light-heartedly, commercial dating. A number of these which are no longer confidential to their formulators are described in matrix form in Figures 5 to 9.

Figure 5(a) in particular is worthy of some extended comment. It was developed by members of the Ministry of Technology's Programmes Analysis Unit. The intention was:

(i) to use morphological matrices to derive, objectively, all the conceptual systems which together comprise marine technology;

(ii) to examine the derived systems, including extant, immediately feasible and speculative systems, for common characteristics. A commonality exists for instance between sonar devices used for fish detection and sub-bottom marine survey instrumentation; then

(iii) to attempt a technological forecast of the ubiquitous characteristics to determine areas of possible economic benefit which could help determine directions for marine technological development.

Figure 5(a). *Morphological Analysis for Marine Communications: Communications-Master Morphological Matrix*

I_d I_a	*Information*; Digital or analog
C_r C_i C_o C_α	*Carrier/transducer*; Radio, infra-red, optical or acoustic
T_r T_i T_o T_α	*Free transmission*; Radio, infra-red, optical or acoustic waves
G_c G_n G_i G_o G_α G_s	*Guided transmission*; Cable, wave-guide, heat-guide, light-guide, sound-guide, and satellite

Source: Garrett, T., 'Illustrations of Technological Forecasting for R & D Evaluations', in Wills, G. S. C. et al., (1969)

Figure 5(a) is the morphological master matrix developed under (i) above for the communications sector of marine technology. Sub-matrices were constructed for each conceptually possible system with each sector. A satellite communication system is hence

(Ia; Cr; Gs; $\check{A}\hat{E}$)

The final notation in this system specification is the environment in which it would be employed. Marine regions were clarified into regions as indicated in Figure 5(b), A to K. $\check{A}$ denotes below boundary A; $\hat{E}$ = above threshold E.

A fishing system was also extracted from its master matrix for fishing for Demersal fish in a tropical ocean as follows:

(Lee + Lse; Hn; Cbt; Wm; Pdf; Bcp; Dpm; Tc; $\check{D}\hat{K}$),

where Lee + Lse = shipborne and externally based fish location equipment

Hn	=	naturally occurring fish shoals
Cbt	=	bottom trawls
Wm	=	mechanical preparation of fish inboard
Pdf	=	deep-freezing preparation
Bcp	=	storage in consumer packs
Dpm	=	waste disposal by conversion to fish meal and pulverization
Tc	=	transport by catcher vessel
$\check{D}\hat{K}$	=	undersea environmental region

The master matrices yielded some twenty systems of marine transport, twenty-two systems of marine communications and many thousand possible fishing systems. Further analysis enabled known systems and potentially viable directions to be identified. (In this way a valuable input for the normative development of scenarios can be developed and we shall look at this further in the next chapter, 2.3.) Nevertheless, once again the analysts reported that they were swamped by the arrayed alternatives and vexed by a particular shortcoming of such analysis – it omits to treat the cost dimension implicit in rates of growth. The exercise was reported as being of considerable value, however, in ordering thought processes and ensuring that important viable alternatives were not overlooked in early stages of development within this area of technology for which so much is forecast in the century ahead of us.

Figure 5(b)

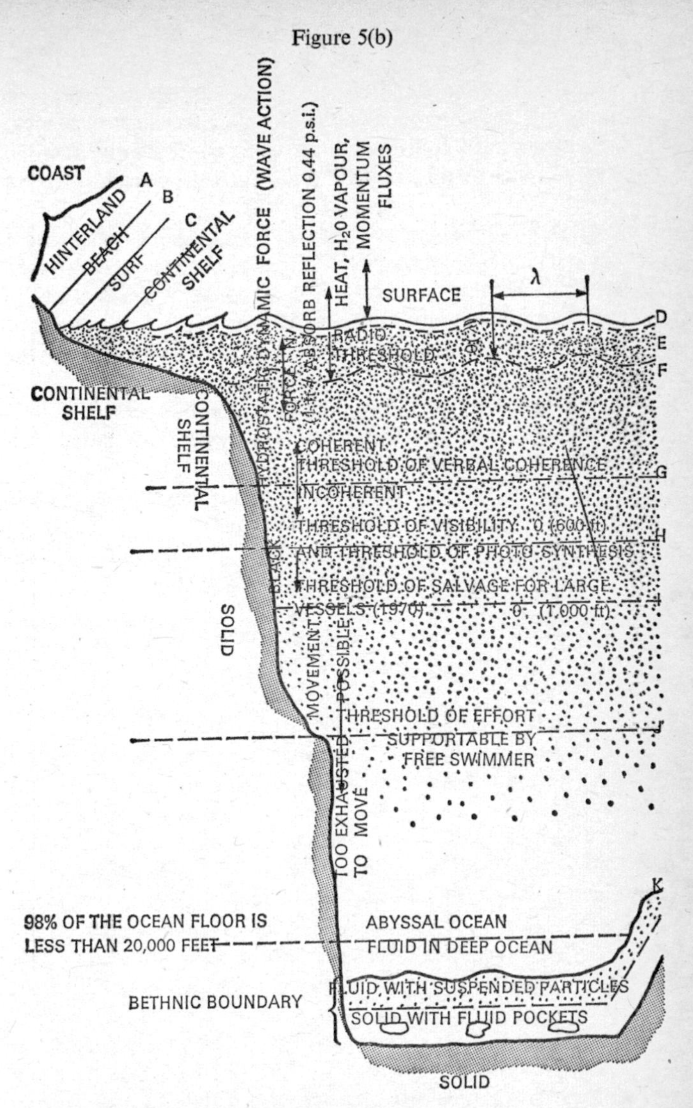

Environmental regions are labelled V and Λ.

Example: DI denotes the region above threshold I and below boundary D.

Source: Medford, R. D., in Arnfield, R. (ed.), (1969)

Figure 6. *Morphological Analysis of Functional Fluids*

	1	2	3	4	5	6	7	8	9	Notes
A Effect	Renders flexible	Reduces friction	Energy transfer Mech Heat Elect Sound/Light	Adhesion	Absorption	Dispersion	Protection anti-rust	—	—	Dispersion here is the ultimate effect not the means as in line E
B Interfaces (macro- or molecular levels)	Metal Metal	Plastics Plastics	Fibres Fibres	Inorganics Inorganics	Elastomerics Elastomerics	Other solids Other solids	Liquid Liquid	Gas Gas	— —	Those are the ultimate interfaces between which the functional permanently resides and achieves the effect on line A
C Initial Inter-facial phase	Solid	Liquid	Gas	—	—	—	—	—	—	This is *the* functional fluid
D Ultimate inter-facial phase	Solid	Liquid	Gas	—	—	—	—	—	—	ditto
E Process used	Temperature change	Pressure change	Reaction	Precipitation	Diffusion	Direct application	Electrostatic	Dispersion	Solution	Means of locating interfacial phase dispersion here is the *means* not the effect as in line A

Known Technologies are:

A1	B2/B2	C2	D2	E1		Plasticization of PVC
A2	B1/B1	C2	D2	E6		Liquid lubrication of metal faces
A3	B1/B1	C3	D3	E3		Internal combustion engine
A1	B1/B1	C2	D3	E1	E2	Steam engine
A4	B4/B4	C2	D1	E8	E3	Cementing bricks
A3	B1/B1	C2	D2	E2		Hydraulic fluid system
A7	B1/B8	C2	D1	E9	E6	Solvent applied rust preventative
A6	B2/B2	C2	D2	E8		PVC Rigisol—Plasticizer is only a processing aid
A5	(B4 or B6)/B8	C2	D2	E5		Air filter medium

A suggestion is:

A1 — B4/B4 — C2 — D2 — E1 Plasticized (non-crystalline) inorganic material

*Prepared privately by H. Jones and Dr J. Grigor.

Figure 7. *Morphological Analysis; Retail Sales Data Capture*

Parameter	1	2	3	4	5	6
A. Data Content	Price	Unit	Size	Colour	Weight	Picture ——→
B. Data Carrier	Package	Tags	Stamped on	Item	Punched card	——→
C. Data Capture Technique	Vending machine	Cash register	Light pen	Scale	TV camera	Optical scanner——→
D. Data Transfer	Wire	Physical	Radio	Audio	————	——→
E. Data Validation	Limit check	Format check	Parity	CDV (check digit verifier)	————	——→
F. Data Processing	On line (real time)	On line	Off line	————	————	——→

Figure 8. *Morphological Analysis: System for Commercial Dating*

Parameter	1	2	3	4	5	6
A. Potential Customers	Male	Female				
B. Customer Derivation	Advertising	Personal selling	Word-of-mouth	—	—	—
C. Matching	Computer	Tabulator	Lottery	Third-party search	Sampling	Selection
D. Output Format	Alpha Numeric	Graphic	Photographic	Voice	Binary	—
E. Output Mechanism	Hard copy Print out	Photo	Microfilm	Magnetic tape	Punched card	Punched tape
F. Communications Medium	GPO mail	Telephone	Personal call	Radio	TV	Satellite

Figure 9. *Morphological analysis for ocean transport*

Parameter	1	2	3	4	5	6
A. Cargo hull stage	Conventional	Rectangular cylinder	Circular cylinder	Sphere	—	—
B. Cargo hull construction	Rigid	Flexible	—	—	—	—
C. Cargo hull units	One	Two	Three	Four	Five	Six
D. Crew housing	Integral	Separate (rigid)	Separate (flexible)	Separate		
E. Crew housing propulsion	Integral	Separate (rigid)	Separate (flexible)	—	—	—
F. Ocean propulsion	Integral	Separate (rigid)	Separate (flexible)	—	—	—
G. Cargo hull position	Fully submerged	Semi-submerged	Clear	—	—	—
H. Crew housing position	Fully submerged	Semi-submerged	Clear	—	—	—
I. Propulsion unit position	Fully submerged	Semi-submerged	Clear	—	—	—
J. Propulsion method cargo	Propeller	Water jet	—	—	—	—

Figure 9 (cont'd). *Morphological Analysis for Ocean Transport*

Parameters	1	2	3	4	5	6
K. Propulsion method crew	Propeller extra	Jet extra	Hull propeller	Hull jet	—	—
L. Support cargo	Displacement	Hydroplanes	Foils	Displacement/ Hydroplanes	Displacement/	—
M. Support crew	Displacement	Hydroplanes	Foils	Displacement/ Hydroplanes	Displacement/ Foils	—
N. Support drive	Displacement	Hydroplanes	Foils	Displacement/ Hydroplanes	Displacement/ Foils	—
O. Speed range	Below convention	Convention	Above convention	High	—	—
P. Information crew/cargo	Direct	Cable	Radio	None	—	—
Q. Information crew/drive	Direct	Cable	Radio	—	—	—
R. Trim control	Displacement	Foil	Displacement/ Foil	Foil/Ballast	—	
S. Linkage C, D, E, F	Rigid	Flexible	Non-applicable	— —	—	—

We have not yet looked very closely at the problems inherent in the delineation of the characteristic parameters and their alternative states, in morphological matrices. We can usefully do this now since this is the aspect with which most learners have greatest difficulty. The most honestly reported struggle with the delineation of characteristic alternative states has been provided by Wills and Hawthorne of Urwick Technology Management Limited. Furthermore, their exposition focuses our attention on relating morphological analysis to the process of new product development within business. They examined a process producing a one-piece metal component. In terms of seven basic parameters a maximum of 5,184 possible combinations emerges. The matrix is shown as Figure 10 (a) and (b).

The process parameter could feasibly be extended also to include

Figure 10(a). *Morphological Matrix; Metalworking a one-piece Component*

Parameter	1	2	3	4
A. PROCESS	Removal	Deformation	Formation	
B. ENERGY	Chemical	Force	Heat	Biological
C. SCALE	Atomic	Molecular	Particulate	Bulk
D. MATERIAL STATE	Gas	Liquid	Solid	
E. ENVIRONMENT	Vacuum	Gas	Liquid	Solid
F. ENERGY SOURCE	Internal to material	External to material	Hybrid	
G. ENERGY-TIME RELATIONSHIP	Steady	Transient	Cyclic	

Figure 10(b). *Morphological Matrix; Metalworking a one-piece Component*

Parameter	1	2	3	4
A. PROCESS	Removal	Deformation	Formation	
B. ENERGY	Chemical	Force	Heat	Biological
C. SCALE	Atomic	Molecular	Particulate	Bulk
D. MATERIAL STATE	Gas	Liquid	Solid	
E. ENVIRONMENT	Vacuum	Gas	Liquid	Solid
F. ENERGY SOURCE	Internal to Material	External to Material	Hybrid	
G. ENERGY-TIME RELATIONSHIP	Steady	Transient	Cyclic	

Source: Wills, R. J., and Hawthorne, E. P., in *Design and Innovation Group Symposium*, (1969)

Deposition and Fabrication, although these are sub-classes of Formation. Whilst fundamentalist analysts would argue that only Formation should appear, for practical purposes a greater number of states can usefully be employed. Over-provision that affords additional analytical insight is justified. A further instance occurs for the energy parameter. Here Biological Energy is shown even though it is a sub-class of Chemical Energy, since it suggested a sufficiently different set of mechanics to the analysts concerned.

Figure 10 indicates two of the systems which were explored – the opposites of one another. The solid boxed system (A1, B4, C2, D3, E3, F3, G1) in Figure 10(a) represents the small scale removal of metal in a liquid medium by steady biological action, whereas the solid boxed system (A3, B4, C2, D3, E3, F3, G1) in Figure 10(b) represents the building up of solid materials from products of biological action.

Other possibilities are: (A1, B1, C1, D2, E3, F2, G1), an (electro-) chemical removal of liquid material suspended in a liquid medium, with a steady externally applied energy source; or (A2, B2, C4, D3, E3, F2, G2/3), which is conventional explosive forming with an alternative energy–time relationship between transient and cyclic. This latter suggestion is valuable especially because of the rule of thumb it employs – take a conventional system and vary just one parameter to see what improvements come to mind. This helps us face up more effectively to the otherwise overwhelming task of threading our way through an evaluation of 5,184 combinations. They have dubbed this method the *small perturbation approach* to morphological matrix evaluation. Equally valuable, if more difficult to visualize, is *morphological mapping in hyperspace* to which we shall shortly turn our attention. First, however, we can usefully examine some of the problems encountered in overlapping morphologies and we shall take examples from architectural design and brewing.

We have frequently emphasized that a delivered satisfaction for an end-customer will almost always be achieved via the aggregation of several fields of technological development. Whilst the metal worked one-piece component just described may well be sold as *a* product to *a* customer it will almost invariably go to form part of a more meaningful form with its own structure within that total system. Trade-offs between elements in its design will frequently be necessary. Hence, it is critical that we recall Zwickey's rule (i) that the problem to be solved must be stated with great precision. There may indeed be elements of a total system which are only partly contributing to the pattern totality and have areas without. Figure 11 demonstrates this point.

Whereas there is no basic reason why all systems A, B, C, D and E cannot be considered simultaneously, it is often found to be effective to treat them only in so far as they impinge on the problem focus – the shaded area.

Poyner has reported an instance of the problems of design or morphological reconciliation in office letting-space in Birmingham. The problem arose when over-provision was made for large letting-areas and they had to be subdivided for smaller tenancies. Ten patterns of possible layout were identified as shown in Figure 12.

These patterns, which are not of course an exclusive listing, form putative elements in the total morphology of any total floor plan. Their interrelationships are critical to any design mix which might materialize. Pattern 1 is highly flexible and unlikely to preclude the use of other patterns in tandem. Conflicts arise, however, between patterns 5 and 6. 5 requires deep plans and 6 limits plan depth. To determine an adequate overall solution these conflicts must be resolved.

Figure 11. *Morphological Methods*

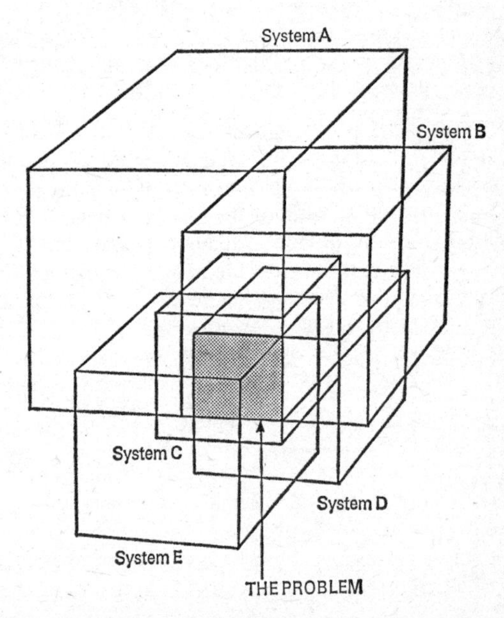

Figure 12. *Ten Patterns for Sub-divided Office Tenancies*

1. Lettable floor space should have simple rectangular plan shapes.
   This pattern facilitates the economic partitioning of space into conveniently shaped small rooms.
2. Floor space arranged in continuous areas of at least 2,500 square feet unbroken by corridors, staircases or service cores.
   This permits variable combinations of tenancy size.
3. Public access space must penetrate the plan so that it gives access to all likely unit areas, say to units as small as 500 square feet.
   Without this pattern it is impossible to give access to small tenancies.
4. Access from lifts to each tenancy should be simple and direct, with as few changes in direction as possible. Maximum distance less than 100 feet.
   Good access is an attractive selling point for small tenants.
5. Deep plan forms, with compact centralized public access space, so that the ratio of lettable space to non-lettable space is as high as possible, say at least 3:1.
   Service space earns no rent, therefore developers try to minimize it.
6. No office space should be further than 38 ft from public access space, and the depth of any office space running parallel to a corridor should not exceed 34 ft.
   This provides adequate escape from individual tenancies.
7. If the public access space referred to in Pattern 6 is not a protected and ventilated lobby to a staircase, then it must be a corridor joining two separate exits to such lobby or staircase.
   The corridor must be no more than 200 feet between exits.
   This provides alternative means of escape from floors.
8. Lettable areas should have compact plan shapes, avoiding long narrow shapes, say not narrower than 1:4.
9. Avoid 15–30 ft depths for office space (that is, the perpendicular distance from the window wall to a corridor or similar limit.)
   This avoids deep narrow rooms and confined inner rooms.
10. Lettable space should be free of columns and other similar obstructions.
    This prevents internal columns limiting the range of possible internal layouts.

Source: Poyner, B., (1969)

Figure 13. *Floor Plan; Conflicting Pattern Relationships*

	1	2	3	4	5	6	7	8	9	10
1										
2			●	●	●		●	●		
3		●		●	●			●	●	
4		●	●			●	●			
5		●	●			●	●	●	●	
6				●	●			●	●	
7		●		●	●			●		
8		●	●		●	●	●		●	
9			●		●	●		●		
10										

Numerical key provided in Figure 12
Source: Poyner, B. (1969)

In Figure 13, the full range of potential conflicts envisaged by Poyner is indicated.

No necessarily optimum solution for the elimination of potential conflicts was identified but, using the morphological analysis shown, the patterns were matched against four major floor-plan forms – core, semi-core, cluster and linear – and the short fall was determined for each. Only the linear plan form offered real scope for modification and this proved analytically possible to accommodate all ten patterns.

Whilst Poyner's analysis is scarcely technological in the sense in which we generally use the term, it is clearly able to demonstrate the aggregative facets of a morphological approach and its identification of conflicts.

Less tractable problems are quite often presented in process technologies such as are found in the brewing industry. The process is conceived of as a series of discrete steps, and whilst at each level analyses are practicable, the state of knowledge at some process levels precludes the full deployment of continuous process production as we know it in the chemical industry at large. The discrete characteristics of the process are described by Royston as:

(i) Development of proteolytic and ameolytic enzymes;
(ii) Preparation of barley starch in a form which can be readily attacked by the various enzymes;
(iii) Reaction of enzymes on barley starch;
(iv) Extraction of fermentable sugars;
(v) Boiling of extract;
(vi) Extraction and isomerization of hop flavour substances;
(vii) Fermentation of fermentable extract by yeast;
(viii) Stabilization of chill haze precursors;
(ix) Clarification;
(x) Carbonation;
(xi) Sterilization.

These constitute a basis both for full-scale morphological analysis to display alternatives and also scope for what we have already described as the small perturbation approach. To most beer drinkers clarity of the precious liquid is a prime requirement. It results from the chill haze stabilization process parameter. A recently advanced realization of the nature of chill haze has led to the development of many processes for its reduction. Filtration is already widely practised as is the addition of proteolytic enzymes to break down the protein fraction of the chill haze precursor, and improved use of proteolytic enzymes is confidently expected.

In both these instances, reported at an excellent conference held at the University of Aston in September 1969 by its Design and Innovation Group, we have seen something less than total employment of morphological analysis. Nevertheless, its orientation has proved of considerable value to the analysis of technological development in both directions.

MORPHOLOGICAL MAPPING IN HYPERSPACE

In any morphological matrix we may develop there will be a number of combinations which have been tried, and directions in which technological development has proceeded historically. The precise configuration of these known and/or viable combinations will have been partly accident of history and partly derived from the extant technological environment when development work first began.

If we can visualize situations in a multi-dimensional way (and many readers may not be able to, since it is a notoriously difficult thing to do), the areas of explored systems for solving a particular problem or delivering a functional capability can be termed intellectually occupied territory. The remainder of our multi-dimensional morphological map is *terra incognita.* Any examination of R & D activity quickly shows that most effort goes into the exploration of the periphery of *terra incognita* where it meets occupied territory.

In any business which commences from a known state of knowledge in a particular field it is quite logical and normal to build on it. We are indeed only seldom capable of any other approach – as Goethe reminded us, 'we can only see what we know'. This sequence is reminiscent of the small perturbation approach. A small contiguous area on the morphological map will be explored. This has been typically seen in the development of combustion engines, with variations in the twin themes of rotary–reciprocating and external–internal combustion. Ayres also points out that, in the development of energy conversion technology, many non-contiguous sectors have been ignored, most noticeably in automotive applications. Largely ignored fields as prime movers include thermoelectric, thermomagnetic and galvanic phenomena.

Morphological space can be sensibly defined as discrete sets of coordinates representing each combination of the parameters and states within them. The space has as many dimensions as there are variables. *Morphological distance* is the number of parameters or states on which any configurations or technology routes differ. An original configuration and one with a small perturbation will be close together in morphological space. A *neighbourhood* is a sub-set

Figure 14. *Breakthrough opportunity index for differing densities of occupied territory on morphological space*

Occupied territory cluster of configurations (max. 27) Column (1)	SURFACE AREA WEIGHTS POSTULATED (normalized *a priori* probability of breakthrough as a function of morphological distance from occupied territory)		
	$\alpha_1=1$; $\alpha_2=1/4$; $\alpha_3=1/9$ etc. Column (*ii*)	$\alpha_1=1$; $\alpha_2=1/2$; $\alpha_3=1/3$ etc. Column (*iii*)	$\alpha_1=1$; $\alpha_2=\alpha_3$ etc $=0$ Column (*iv*)
1	9·89	14·67	6
4	14·36	17·33	12
6	16·50	18	15
9	18	18	18
12	15	15	15

Source: adapted from Ayres, R. U. (1969), p. 84

of configurations such as we find under the small perturbation approach, and such as we expect to find within given fields of development within a technology, e.g., internal combustion engines. The *surface* of a neighbourhood is the set of all configurations which differ at most in a single parameter, and *surface area* is the sum of all such points. A *weighted surface area* is defined as the sum of the points differing by one, two, three, etc. parameters times an appropriately decreasing coefficient, α_1, α_2, α_3, etc.

A *technological breakthrough* can now be identified as a newly actualized configuration but not as improvements within an existing actualized configuration.

The purpose of this definitional preamble is to explore the probabilities of meaningful breakthroughs and thereby to guide the screening of the abundant alternatives generated by morphological analysis. *The probability of a technological breakthrough, per unit of time, will be a decreasing function of its morphological distance from occupied territory, other things being equal.* Quite what function we should hypothesize is hard to tell, but an inverse quadratic has been suggested by Ayres implying a normalized *a priori* probability for distances of 1, 2, 3, 4, etc. of 1, 1/4, 1/9, 1/16, etc.

It can further be deduced that *the opportunities available in any given morphological surface area at any point in time will be approximately proportional to the surface of the weighted area corresponding to the surface of the occupied territory.* If we assume an inverse quadratic functional relationship, with α_1, α_2, α_3 weights as 1, 1/4, 1/9, etc., we can perceive that opportunities for technological breakthroughs are smallest at start and finish of the exploration of morphological space. That is to say, when there is just one configuration or a high saturation of morphological space.

In Figure 14 we have assumed 27 combinations of a 3-parameter technology. Clusters differing on 1, 2 and 3 parameters can be identified if we assume a single occupied territory combination. There will be 6, 12 and 8, respectively. Occupied territory of 6 combinations would provide 15 configurations differing in one parameter, 6 in two and none in three, and so forth.*

Column (ii) has applied surface area weights in line with the inverse quadratic function already discussed for technological breakthrough probabilities in relation to morphological distance. Two other weighting systems are employed in columns (iii) and (iv) with similar effect. Maximum opportunity within this technology sector occurs between the 6 and 9 clusters of occupied-territory configurations.

The enumeration within any particular technology field of occupied territory is not an impossible task and it facilitates the approaches we can make to the screening of the myriad alternatives we generate. The *breakthrough opportunity index* can also be expected to constitute a valuable guide in the allocation of R & D appropriations to differing technology routes and in the forecasting of switchover points in the deployment of given technologies to meet functional capabilities. This is an aspect we have already considered in trend extrapolation in Chapter 2.1.

* The full array used in Table 14 is, for each clustersize of occupied territory, 1, 4, 6, 9 and 12, for configurations differing on 1, 2 and 3 parameters respectively, as follows: **1** – 6, 12, 8; **4** – 12, 9, 2; **6** – 15, 6, 0; **9** – 18, 0, 0; **12** – 15, 0, 0.

If morphological mapping enables R & D effort to be related at least at the macro-level to breakthrough opportunities, we have so far failed to examine the relating problem for the firm or the market-place directly. I, as a marketeer, am sad to say marketing has been a gravely neglected area of analysis. None the less, the potential for matching customer-specified needs with morphological analysis at the macro-level seems to exist in certain marketing research procedures – most noticeably the family of techniques which undertake what is termed *product gap analysis*. This in itself is a morphological method but is not based on the objective specification of product parameters. Rather, customers are invited to indicate what they consider to be the relevant parameters for adjudging market offerings and using principal component analysis these may often be reduced to some five or six. We can now once again revert to our multi-dimensional perception procedures and will realize that at any given time a range of products offered on a market can be subjectively categorized as occupied territory within morphological space on those principal parameters. Customers can identify just exactly where all extant market types of a product fall and *terra incognita* is the remainder. Much of that will be unfeasible or improbable territory, or make mutually incompatible demands in the light of known technologies. Nevertheless, if a concept of an ideal product or ideal offering to satisfy functional need can be elicited from representative customers, the shortfall of existing offerings can be identified. This is precisely the process we saw at work for the sub-division of office space for small tenancies earlier on.

Once functional capability shortfall has been determined against the customer assessment of the ideal, a *market desirability weighting* can be added to the breakthrough opportunity index we earlier described. That index was, of course, solely a feasibility assessment of the situation and to decide research direction on that alone is to a marketeer a deadly sin.

The concept of the ideal is not difficult to envisage nor to determine although it is continually changing in the light of customer awareness. It is also subject to the danger of technological myopia, since customers cannot ask for what they do not know nor even question what they are totally accustomed to expect. This latter point is well illustrated by the shock and amazement with which the market leaders in the paint industry saw thixotropic (non-drip) paints sweep the home-decoration market in the early 1960s. No indication had come from customers that they were dissatisfied with paint which dripped even when questioned about disadvantages of the product. Dripping was accepted and perceived as inherent in the product. The launching of thixotropic paints, none the less, revised that pattern of expectations dramatically within that particular market segment.

If customer specification of an ideal market offering is at times a fallacious approach, it can and has been used judiciously by many. It can act in this context as an important weighting element within the total screening process for the alternatives.

In these comments on gap-analysis procedures from the marketing end of new-product development, I have been describing in part what the technologist terms *system sensitivity*. It proceeds from an analysis of existing systems available on a market to provide functional capability and seeks to determine which parameters are critical to the field of use. Wills and Hawthorne, whose morphological analysis we quoted in Figure 10, identified within the metal-working area those aspects of the task to which a successful batch-work machine must be insensitive if it is to be a market success. (These have equivalence with the ideas we mentioned earlier from customer-derived analysis of extant market offerings). They listed:

(i) utilization
(ii) product geometry, material, accuracy, batch and total quantities
(iii) operator error
(iv) quality of shop management

If systems are not insensitive to these elements, their superiority will be lessened in many applications. Numerically controlled machine tools are in this category and have diffused slowly. The product cost reductions of 50–75 per cent on paper are easily dissipated by variation in utilization and product mix.

These concluding comments emphasize the vital need to incorporate market screens in the evaluation of morphologically generated alternatives at the earliest possible date, treating them sometimes as constraints but more often as an intelligent navigational device. A delivered functional capability must be a system that is readily marketable at a profit, not something beautiful based on technological feasibility alone.

CHAPTER 2.3

Normative Approaches and Normex Reconciliation

NORMATIVE approaches to TF begin with a statement of a need, and then seek to identify how it might be achieved. This is quite the reverse of the processes we have indicated in Chapters 2.1 and 2.2, where forecasting was based on the exploration of possible technological futures from a present knowledge base. That a reconciliation must take place between any normative forecast and the current state of knowledge is readily apparent. We shall look later at a particular process of reconciliation dubbed *normex* (normative extrapolative).

First, however, let us examine the major normative techniques – scenario iteration, impact analysis, and relevance-tree analysis/contextual mapping – and see for each how it contributes to an approach to TF which has only become possible in the last decade or so. Figure 1 demonstrates the distinctive character of normative approaches. In particular, that paradigm emphasizes man's ability, if he so desires, to plan to avoid the effect which extrapolated trends would occasion.

Figure 1. *The Normative Approach in TF*

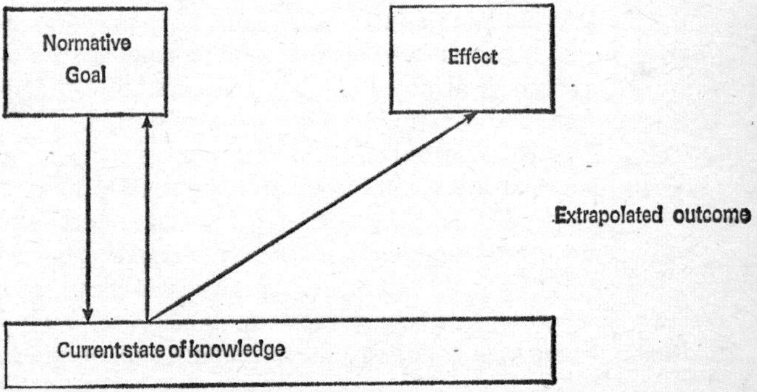

Normative Goal

Effect

Extrapolated outcome

Current state of knowledge

SCENARIOS – ANTICIPATIONS – FUTURIBLES

Scenarios, anticipations, futuribles, even futures, are terms frequently employed to identify the normative goals which any social or corporate system may set itself. The normative approach is not uncommon in business planning, where a company seeks to change direction on the basis of its understanding of the direction in which it would otherwise move. At the technological level, however, such anticipatory behaviour has hitherto been regarded as the preserve of science-fiction writers, of futurists. Along with this new trend has come the inescapable need to describe normative social and economic goals and attitudes which we may seek to attain or engender, since these constitute the environment in which our technological offering must find its effective market.

Denis Gabor, who in 1964 wrote *Inventing the Future*, a book which made a very wide impact, began a movement amongst scientists for a more selective approach towards possible futures. 'Till now man has been up against Nature; from now he will be up against his own nature,' he commented. A future can be developed of which Mozart would be proud which avoids three major dangers of contemporary civilization – nuclear war, an age of leisure, and over-population. A scenario was written for the identification and development of a society where man can not only survive but enjoy his life.

Such attitudes are relatively recent. It is a fairly new phenomenon to find a general (not specific) recognition of science's responsibility to societies and nations; a realistic understanding of economic and technological potentials; an awareness of the constraints; and genuine ability to hedge in the face of threats.

McLuhan has, with great panache, identified us as standing on the threshold of a new age, not of mass communications, but of person-to-person potential. The Biafran and Vietnam wars have shown us just how communications can personalize a struggle and transform attitudes and actions from what they were thirty years ago in earlier conflicts. The same is true of the influence of communications on social, economic and political matters, and, of course, on technology.

Normative forecasting, however, can only be meaningful

(i) if the levels to which it is applied are characterized by constraints. Normative forecasting can be applied to the levels at which events impact on one another only if they are *sufficiently* closed by natural or artificial forces, or by consensus or values;

(ii) if the number of opportunities which exist, and are recognized to be on these levels, are greater than the number we can pursue under contemporary constraints. Only in such a state of imbalance can selection between alternatives be made.

Scenario writing as a specific technique attempts to set up a logical sequence of events, in order to demonstrate how a given normative goal might be achieved. A time-frame is often specifically introduced for the process of iteration, but as we shall see in the next chapter, on Delphi time-scaling, it can be suggested on a carefully determined basis.

The most fruitful environment for the development of valuable scenarios is naturally the most creative. Considerable imagination is needed, and when formally undertaken a wide variety of social psychological techniques are employed. The source of any particular scenario is not sacred, so it is customary to spread the net as widely as possible throughout an industry to intermediate manufacturers or distributors and to end-users or customers of a product or process.

The elaboration of scenarios enables a feeling for critical decision points to emerge when a specific problem area is handled. Scenario writing for long range planning has been pioneered by Herman Kahn, particularly in his book *The Year 2000*, which he wrote with Wiener as a framework for speculation. He cites two main advantages for such an approach:

(i) scenarios lessen carry-over thinking; they force one to plunge into the unfamiliar aspects of the environment by dramatizing the possibilities on which they focus;

(ii) scenarios force the analyst to examine the dynamics of situations which might otherwise be overlooked.

Figure 2 indicates a Kahn scenario for the post-industrial society. Such procedures are employed by a number of large companies and have proved especially valuable in opening up new directions for R & D. Unilever, through Ronald Brech, made public its scenarios for 1984 in *Britain 1984: Unilever's Forecast*. They explored the likely social development, changing patterns of technology, psychological and economic factors, politics and demography as the background for Unilever's long-range operations. The procedures used contained less recognition of normative

Figure 2. *Kahn's Scenario for the Post-Industrial Society*

1. *Per capita* income about fifty times the pre-industrial
2. Most 'economic' activities are tertiary and quaternary (service-oriented) rather than primary or secondary (production-oriented)
3. Business firms no longer the major source of innovation
4. There may be more 'consentives' (*vs.* 'marketives')
5. Effective floor on income and welfare
6. 'Efficiency' no longer primary
7. Market plays diminished role compared to public sector and 'social accounts'
8. Widespread 'cybernation'
9. 'Small world'
10. Typical 'doubling time' between three and thirty years
11. Learning society
12. Rapid improvement in educational institutions and techniques
13. Erosion (in middle class) of work-oriented, achievement-oriented, advancement-oriented values
14. Erosion of 'national interest' values?
15. Sensate, secular, humanist, perhaps self-indulgent criteria become central

Source: Kahn, H., and Wiener, A., (1967), p. 186

criteria than we might have hoped for, but the study was a pioneering venture. I have similarly written an examination of marketing futures, attempting to identify the process whereby a normative, desirable (to me), pattern of Mozartian marketing activity can occur in the post-industrial society towards which we are undoubtedly moving. The scenarios were written in the context of an evaluation of the organizational needs for marketing in UK industry in the next two decades as part of a study sponsored by the British Institute of Management. These 'futures' and the full study results appear in R. Hayhurst and G. Wills, *Organizational Design for Marketing Futures.*

Lord Keynes developed a vitally important scenario for the post-industrial society in 1930 under the appropriate title *Economic Possibilities for our Grandchildren*, reprinted in *Essays in Persuasion* by J. M. Keynes in 1963. He considered the problems raised by the accumulation of wealth through investment, the danger that success breeds failure rather than generosity, humaneness and integrity. He wrote:

> The economic problem, the struggle for subsistence, always has been hitherto the primary, most pressing problem of the human race. If the economic problem is solved, mankind will be deprived of its traditional purpose.
>
> Will this be of a benefit? If one believes at all in the real values of life, the prospect at least opens up the possibility of benefit. Yet I think with dread of the readjustment of the habits and instincts of the ordinary man, bred into him for countless generations, which he may be asked to discard within a few decades. . . . Thus for the first time since his creation man will be faced with his real, his permanent problem – how to use his freedom from pressing economic cares, how to occupy his leisure, which science and compound interest will have won for him, to live wisely and agreeably and well.

The specific device of *gaming* is often used to assist in the iterative process. Several teams are situated within a problem sector and begin to postulate courses of action. The varying decision sequences in which each team becomes engaged forces scenarios forward. War games and diplomatic thrust and parry routines have been most frequently played out in this way, but increasing use is made of the approach in technology sectors. It acts as a useful supportive method for brainstorming/think-tank efforts, and for the careful structuring of organizations to maximize the germination, gestation and diffusion of new ideas.

[Readers who are mystified as to how a scenario can be sufficiently simplified to enable quantification of the estimates are asked to wait until Chapter 2.4. There, simply stated scenarios, e.g., 50 per cent of all clothing sales in moulded plastic, are the input to the Delphi method.]

IMPACT ANALYSIS

We have spoken of the impact of technological advance of events which are certainly highly likely to occur unless some hedging action is undertaken and unless some normatively derived goal is focused upon instead. The process of impact analysis provides extensive iteration of the full range of influences a given event will have. There are perhaps no two more widely influential technological phenomena in society today than the computer and automation. These two topics have obsessed the forecaster time and again, and have given rise to endless predictions of the likely impact on various sectors.

Ozbekhan reported in 1967 on the outcome of a major impact analysis of automation wherein he postulated six technological advances as follows:

(i) By 1975 computer hardware capability will become at least half as expensive as at present; software will decline in cost at a far slower rate.

(ii) Hardware will become much more powerful than today in every sense – absolutely, relative to size, relative to cost and to the ability of system designers to use its capabilities.

(iii) Computer design will achieve far more diversity and modularity in systems, permitting much greater flexibility in assembling them.

(iv) More powerful, flexible and inexpensive man/machine interface hardware and software will become available.

(v) Time-sharing technology will evolve into the linking of several or many time-shared (and non time-shared) computers in networks tied together by common carrier communication channels.

(vi) Processor costs will decrease drastically relative to common carrier communication costs.

In Figure 3, the possible applications which would follow such technological advance are indicated. Its importance to managements, and to society, is that it extends the detailed understanding of possible future events and once again enables us, should we so wish, to avoid or mitigate some of the effects.

Such approaches are of value to the smaller and medium-sized firm in a technology-based industry, since they will often afford the opportunity to appraise the impact of achievements by market/technology leaders on their own sphere of operations. Typical examples can be cited in laser technology or holography – both major new technologies whose impact will spread in many directions. The hologram has made three-dimensional recording and projection a practical reality. There was a time until recently when it was thought that laser technology was indispensable to holography, but this is no longer so. The development of filtration of light, image separation, and the subsequent creation of interference by converging the two similar wavelength images has demanded a major reconsideration of impact patterns. Although the holographic applications which have excited most attention are in the realms of three-dimensional television and films, a wide range of other sectors must be prepared for impacts. Among the applications of holography proposed more recently or under development are the following:

(i) Radar systems enabling an airport traffic controller to look into a three-dimensional scope and watch all aircraft in the area.

(ii) Scanners that can map distant planets or enable engineers to see stress patterns in a whirring propeller.

(iii) Computers that display their solution of engineering problems as three-dimensional images. The object that is displayed can be examined from one side or the other, as if it were really there.

(iv) Photographs of fog or dust clouds that can be projected into space in three-dimensional form so that each particle can be examined, classified and counted by microscope.

(v) Machines that can process hundreds of thousands of pictures, picking out rapidly all patterns that conform to

Figure 3 Impact forecast based on Ozbekhan's six postulated technological advances

Main expected outcomes	*probable date*	*probable obstacles*
Planning		
long range problem solving in all areas of application (i)	1980–2000	problems of representation of meaning in mechanically interpretable forms
aids for sensing and understanding environments (ii)		
feasibility testing of long range goals (iii)	1975–1980	
Manufacturing		
replacement of sensory control of workers by more sensitive electro-mechanical devices (i)	1966—	
self-adaptive inventory, production and organization control (i)	1975–1980	
multi- or general-purpose regional (then global) interlinked tools (ii)	unlikely	transportation costs high; unit process costs declining
robots – for production and services (i)	1970–1985	
Exchange		
general transaction facilitation and handling using new unit values (i)	1990–2000	
credit systems, invoice paying, purchasing, international business transactions particularly at commercial level (ii)	1975–1980	
automated marketing through home terminals and touch-tone telephone systems (iii)	1975–1980	cost of delivery of purchases; communication costs vs. increased leisure
Negotiation		
bargaining: automation of the search for meanings and of the determination of real issues in dispute	1975–1985	mechanical extraction of meaning from unstructured information
real issue identification and conflict identification	1975–1985	
Information processing		

all areas of application	1980–1995	‘computer’ utility unlikely as processer costs declining much faster than communication costs; also social problems of privacy
computer/information ‘utility’ – inter-linked data banks, automation of office and institutional data handling	1970–1975	
automated diagnostics (medical, social, mechanical, etc.)	1980–1990	
Learning/Teaching		
regional and/or global educational centres accessible through home terminals (ii)	1985–	global or large regional centres unlikely because of high communications costs vs. low processer costs
substantive curriculum development (iii)	1975–1980	
generalized computer assisted instruction (i and iii)	1970–1975	
man-computer symbiosis (i, ii and iii)	1985–	
knowledge development (i)	1985–	problems of representation
Traffic-communications		
transmission of all communications (i)	ongoing	
high speed message switching and handling utility (ii)	1990–	installation of all-digital communication network.
off-line experimentation and testing of networks (iii)	1960–1975	
Resource management		
complex scheduling systems (i)	1970–1980	
optimal benefit resource application decisions (ii)	1980–2000	
Traffic-transportation		
mass rapid transit (i)	1975–1985	high speed physical object switching
car traffic control (ii)	1970–1980	
integrated traffic system problem solving development (iii)	1975–1980	
Library		
home access and retrieval (reproducing copy)	1985–	communication costs vs. printing costs
text abstraction and reproduction	1975–1980	
question answering	1975–1980	mechanical interpretation of questions to extract meaning
automatic bibliography generation	1970–1975	

Source: *Science Journal*, October 1967

certain criteria. Such devices are already being used to find the signatures of oil-bearing geologic formations from explosion soundings of the earth's interior. The same system can be used to screen electrocardiograms and fingerprint files or search aerial photographs for missile sites. (In other words, the addition of holography to planned uses of high-speed computer evaluation of large volumes of photography, adding increased information capacities from three-dimensional effects.)

(vi) Side-looking radars that enable aircraft flying offshore to map in detail cloud-covered installations along a coastline.

(vii) Microscopes that can display directly the three-dimensional structure of proteins and other complex molecules formed from millions of atoms. Such a capability would revolutionize the development of new drugs.

Source: *New York Times*, 19 March, 1967, cited in Kahn, H. and Wiener, A. J., (1967), p. 105

Holography was, incidentally, discovered by a process popularly known as serendipity, whilst in search of image-magnification techniques, i.e., whilst searching for something quite different. This is a not unfamiliar phenomenon in technological and scientific discovery. Fleming discovered penicillin under similar serendipitous circumstances. The term was first coined by Horace Walpole in 1754 in *The Three Princes of Serendip*, Serendip being a location in Ceylon, where the heroes of this tale make such discoveries. A milder contemporary version in R & D environments is to tackle something quite different when a block seems to be occurring along a particular line of thought. When a later, refreshed return is made to the problem a solution will frequently emerge. Such an approach to creativity and its stimulation should not be confused with analogous thinking such as the application of Darwinian theories of selection to R & D projects or market offerings. Rather, serendipity achieves results since the starting point for consideration is changed and new avenues of development open up in a way which the previously fixed directional thinking precluded.

We have already studied the perceived impact of several facets of future technology in Chapter 1.2, Figures 2 and 3. There, Oto Sulc's questionnaire study, albeit with a group of post-experience management students at the Manchester Business School, examined the relationships of technological and social change. His study has pioneered a new methodology which we can expect to extend to many further sectors. Hall at ICL has described from a societal viewpoint the levels at which the impact of the computer can expect to be felt, as is demonstrated in Figure 4.

Figure 4. *The Social Impact of the Computer*

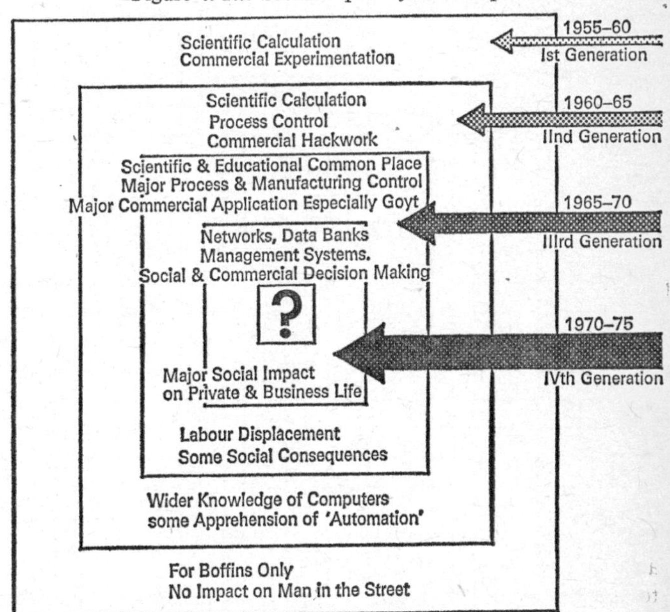

Source: Hall, P. D., in Wills, G. S. C. et al., (1969), p. 194

RELEVANCE TREES AND CONTEXTUAL MAPPING

Relevance trees are a mature development of a fairly direct check-list approach to the inter-relationships of elements in the development of an aggregated performance level. Figure 5, which was

Figure 5. *Relevance Matrix for Docks, Ports and Harbours*

	ITEM	1	2	3	4	5	6	7	8	9	10	11	12	13	14	15	16	17	18	19	20	------	65	66
1	TIDES			✓	✓										✓	✓								
2	TOPOGRAPHY	✓		✓	✓			✓	✓						✓	✓	✓							
3	RIVERS				✓							✓			✓									
4	CURRENTS		✓	✓		✓									✓	✓								✓
5	WEATHER	✓			✓			✓	✓						✓	✓	✓		✓		✓			
6	HINTERLAND									✓	✓	✓												
7	LAND TRANSPORT														✓	✓								
8	AIR TRANSPORT														✓	✓	✓		✓					
9	LABOUR RESOURCES										✓	✓		✓										
10	SHIP REP. & MAINT.																							
11	IND. SERVICES										✓													
12	SOCIOLOGY														✓									
13	ECONOMIC HISTORY						✓	✓	✓	✓	✓	✓	✓		✓									
14	AMENITY		✓	✓								✓												
15	TRAFFIC CONTROL																✓	✓	✓					
16	NAVIGAT. AIDS															✓		✓	✓					
17	BUOYS					✓											✓		✓					
18	RADIO & RADAR		✓													✓		✓						
19	POSITION FINDING		✓	✓	✓										✓	✓	✓	✓						
20	COMMUNICATIONS																							
	¦																							
	¦																							
	¦																							
	¦																							
65	CORROSION										✓							✓						
66	BARRAGES	✓	✓	✓	✓		✓	✓					✓	✓	✓	✓								

Source: T. Garrett, in Wills, G. S. C. et al., (1969), p. 236

developed by members of the Programmes Analysis Unit at the Ministry of Technology charts areas where commonality exists between sectors. Matrices of 66 × 66, which the ports, docks and harbours example come to, are extremely difficult to evaluate, to say the least. To identify priority areas and their sequential needs for resources, a quantitative assessment must be made of each paired relationship, either rationally or subjectively. Such analysis

Figure 6. *Relevance Pattern for Machine Tools*

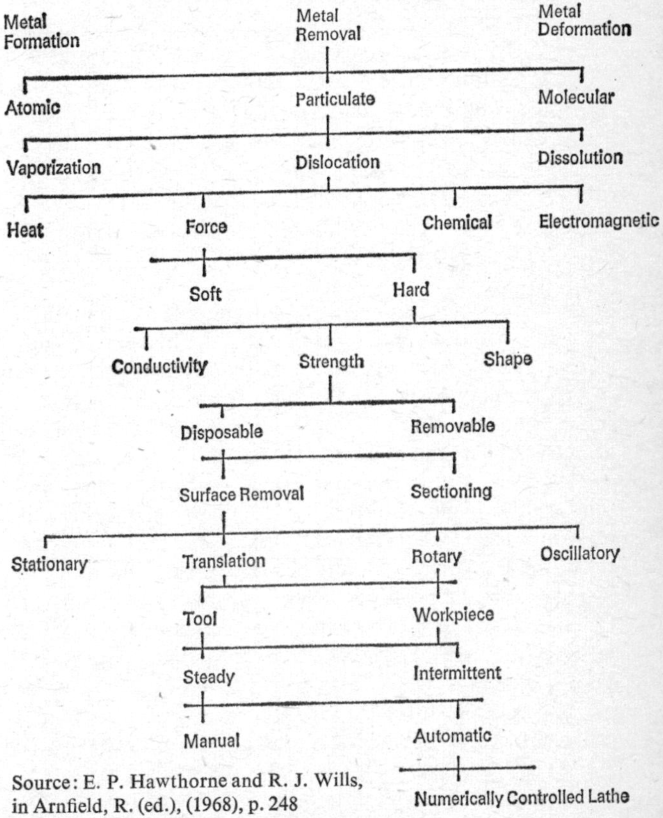

Metal Formation — Metal Removal — Metal Deformation

Atomic — Particulate — Molecular

Vaporization — Dislocation — Dissolution

Heat — Force — Chemical — Electromagnetic

Soft — Hard

Conductivity — Strength — Shape

Disposable — Removable

Surface Removal — Sectioning

Stationary — Translation — Rotary — Oscillatory

Tool — Workpiece

Steady — Intermittent

Manual — Automatic

Numerically Controlled Lathe

Source: E. P. Hawthorne and R. J. Wills, in Arnfield, R. (ed.), (1968), p. 248

scarcely helps to illuminate rates of change inherent in the contributing technologies, but it is a valuable technique for the structuring of a problem.

Figures 6 and 7 demonstrate elementary uses of a tree presentation of relevance developed by Hawthorne and Wills of Urwick Technology in the area of machine tools; and at the Institut Battelle by Harvey Newland for petrol engines with afterburners.

Figure 6 traces the pattern of relevance for a numerically controlled lathe which uses throw-away tools.

Figure 7's analysis begins with the environmental factors that have an influence on energy use and then develops to indicate relationships to transportation. The basic form of energy considered is petroleum, but it could equally well have been electric battery or fuel cell. The path shows that it may be desirable to develop processes for raising the octane numbers of petrol so that they will be satisfactory for cars equipped with afterburner for the suppression of exhaust emissions. These afterburners will not function with fuels containing lead additives for raising the octane rating.

Relevance tree analysis integrates both the tree-like pattern of depiction and matrix analysis at each level. The pioneering study was accomplished by Honeywell, under the acronymistic title of PATTERN (Planning Assistance Through Technical Evaluation of Relevance Numbers). In line with the emphasis of this book on non-military application, however, we shall look particularly at a later application, MEDICINE (Medicated Instrumentation and Control Identified and Numerically Evaluated) and at a non-military application from NASA's Apollo programme, both also engineered by Honeywell. Although our own Ministry of Technology has reported a proposed matrix quantification procedure, no UK commercial applications have yet been published.

The Honeywell technique is conducted as follows. A qualitative scenario is prepared which assesses overall objectives, tasks, approaches, systems, sub-systems, functional elements, technological deficiencies, etc. These constitute the levels of the relevance tree. Figures 8 and 9 show these trees for NASA's Apollo Payload Evaluation and MEDICINE respectively.

Figure 7. *Qualitative Relevance Tree for Petrol Engine with Afterburner*

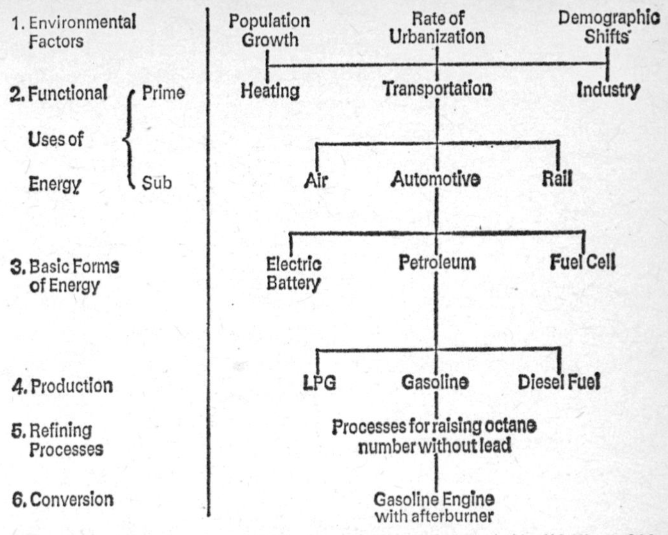

Source: H. Harvey and E. V. Newland, in Arnfield, R. (ed.), (1968), p. 300

Figure 8. *NASA's Apollo Payload Evaluation Relevance Tree*

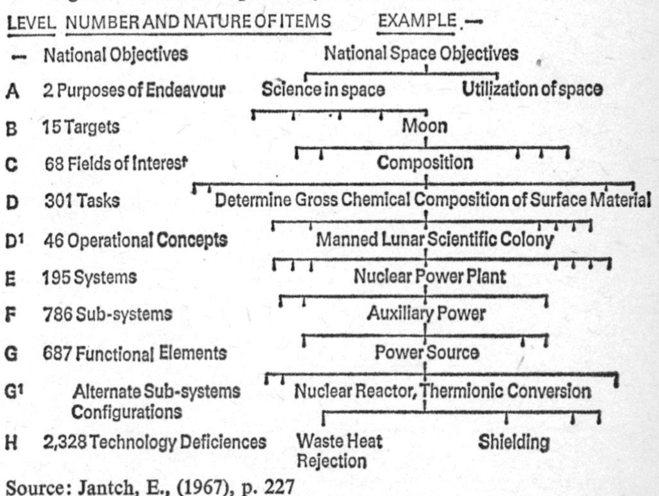

Source: Jantch, E., (1967), p. 227

A procedure must then be followed for the assignment of significance numbers. Technology forecasts are made at primary-systems levels and all lower levels using any and every technique

Figure 9. *Honeywell's MEDICINE Relevance Tree*

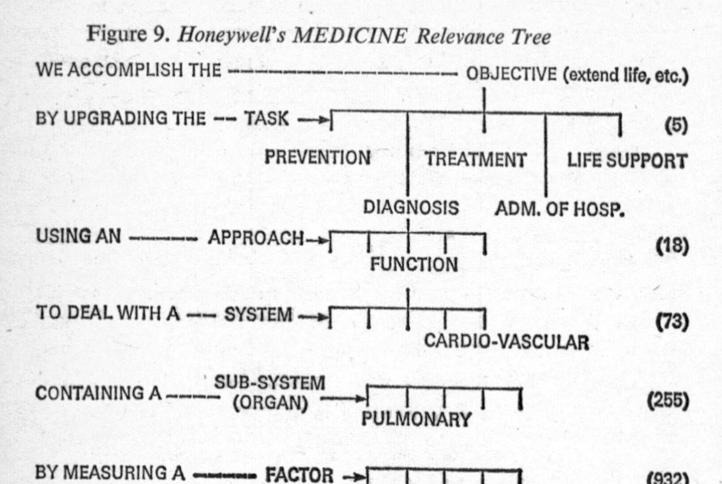

Source: Esch., M. E., (1965), PATTERN; reprinted in Arnfield, R. (ed.), (1968), p. 209

Figure 10. *Paradigm of the Honeywell Relevance Tree Analysis Procedure*

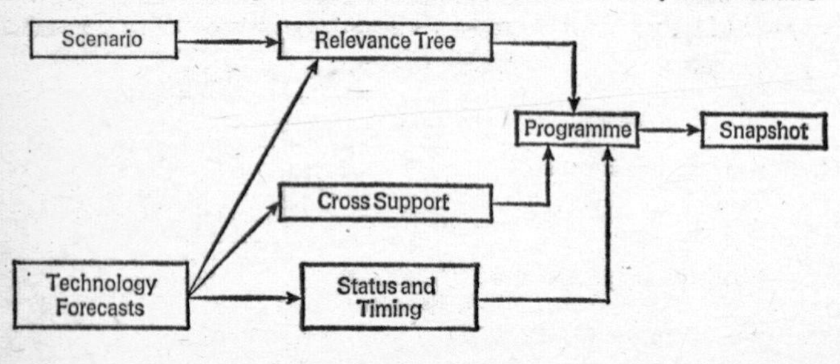

in the book, but particularly the extrapolative techniques described in Chapter 2.1. Apart from these assessments, two other sets of characteristics are examined – *cross support* or spin-off to other areas of advances in given segments; and *status and timing* of systems and sub-systems. This latter set of characteristics is normally categorized under research, exploratory development, advanced development, product design or availability. The paradigm of the whole sequence is given in Figure 10.

Levels within a relevance tree describe the stages of transfer of technology between society, technologies, and their input resources. At each level a matrix is established to match issues against criteria. The matrix takes the general form given in Figure 11. The criteria, their weights and their significance number are estimated by experts on the basis of the scenario. This is, needless to say, a massive task. Two normalizing conditions are introduced. Both the sum of all the criteria weights, ($W_\alpha + W_\beta \ldots + W_\nu$), and the sum of significance number against each individual criterion across all items, ($S^\alpha_A + S^\alpha_B \ldots S^\alpha_N$) are both equal to 1.

The relevance number is then defined, for the Jth item on the *i*th level for criteria, *x*, as follows:

$$R^J_i = \sum_{x=\alpha}^{\nu} Wx\, S^x_J$$

which in turn implies that the sum of all relevance numbers is normed to unity.

The total relevance figure for any particular issue can now be obtained by multiplying upwards to the top of the tree, *a*, from the level in question, *f*. In formula form this is written as

$$\text{Total relevance} = \bigwedge_{i=a}^{f} R_i$$

This, of course, requires refinement for the characteristics of cross support and status and timing of development work. In operational terms these refinements are seldom included but there is no reason why they should not be. It is also possible to incorpor-

Figure 11. *General Form of Honeywell Matrix Showing Inter-related Significance and Relevance Numbers*

Criteria		*Items on tree level i*				
Description	*Weight*	A	B	C	D	.... N
α	W_α	S_A^α	S_B^α	S_C^α	S_D^α	S_N^α
β	W_β	S_A^β	S_B^β	S_C^β	S_D^β	S_N^β
γ	W_γ	S_A^γ	S_B^γ	S_C^γ	S_D^γ	S_N^γ
•	•					
•	•					
•	•					
•	•					
•	•					
•	•					
ν	W_ν	S_A^ν	S_B^ν	S_C^ν	S_D^ν	S_N^ν
Relevance Number at level i		R_i^A	R_i^B	R_i^C	R_i^D	R_i^N

ate not just single estimates of significance and criteria weights but probabilistic distributions. The pattern is extremely flexible.

The most valuable outputs so far from this sort of analysis fall under four main headings:

(i) it enables the upgrading of R & D efforts in relevant mission areas;

(ii) it identifies technological deficiencies in relation to systems and concepts, and indicates their relative importance;

(iii) it facilitates the relative valuation of selected technology improvements in a given area – greater accuracy, lower cost, lower weight, etc.;

(iv) it permits the detailed evaluation of alternatives in the light of the main objectives.

The title to this section included the phrase *contextual mapping*, but so far the discussion has been of relevance matrices and tree analysis. Contextual mapping is a natural by-product of relevance analysis of the variety described. It enables the business to identify precisely where it and the current state of development stand in relation to a particular technological objective. This becomes of particular value in the timing of R & D work. It is reported to have played a decisive role in Fairchild's timing of entry into integrated circuitry. Fairchild waited until a clearly defined and recognizable set of factors showed signs of becoming feasible – especially planer techniques, the concept of isolating p–n junctions and experience in mass-production yield. A strong R & D programme at this later date gained quick results. Contextual mapping relies particularly on the refinements to significance numbers and the resultant relevance number brought about by an understanding of cross supports as well as status and timing.

NORMEX RECONCILIATION

The process of normex reconciliation forecasting is one of the most recently introduced techniques of TF, and few published studies exist outside of the military field. A. W. Blackman, of United Aircraft Research Laboratories, has however reported on such an analysis for commercial jet-aircraft engines and his forecast will be described here in some detail.

The technique's major advantage is that it integrates normative scenarios about future product design with the extrapolation of technical capabilities. It thereby facilitates the establishment of R & D objectives which will be necessary to meet the market requirements at a specified period in the future. It can also indicate the market-oriented trade-offs between various technological parameters by demonstrating the desired interface between technological capabilities and market demand. Finally, as we shall see in detail shortly, the normex technique provides a measure both of the mean and the variance of technological parameter forecasts thereby permitting an evaluation of the uncertainty associated with such forecasts.

The normex reconciliation process employs five analytical stages.

(i) Collect historical data on total annual market sales of the system under examination. (Here we are looking at commercial jet aircraft engines.)

Figure 12. *Specific Weight Variations among Commercial Installations in 1964*

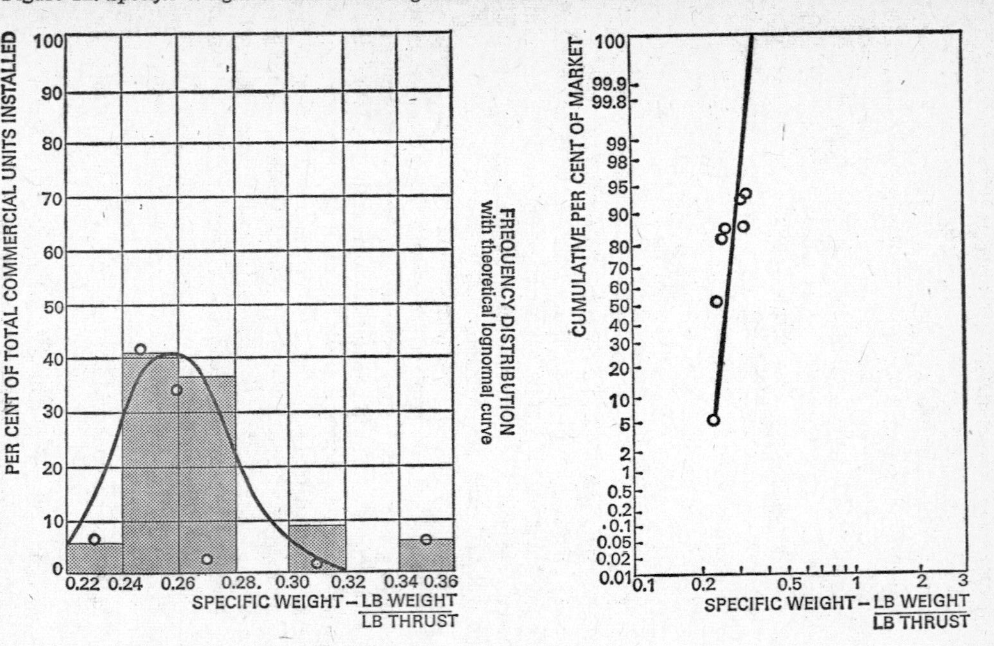

Source: A. W. Blackman, United Aircraft Research Laboratories

Figure 13. *Specific Fuel-consumption Variations Among Commercial Installations in 1968*

CUMULATIVE FREQUENCY DISTRIBUTION

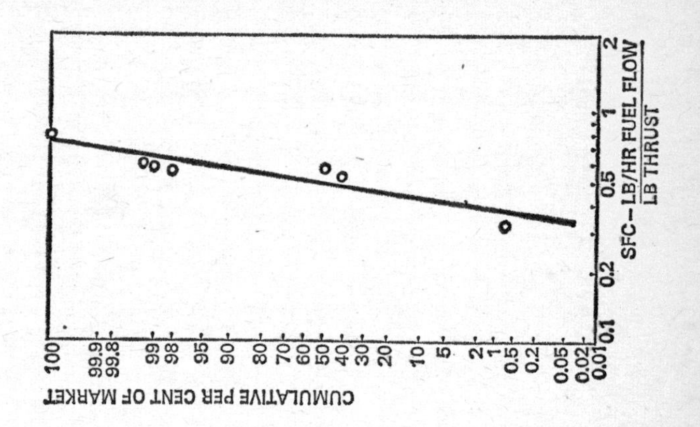

FREQUENCY DISTRIBUTION
with theoretical lognormal curve

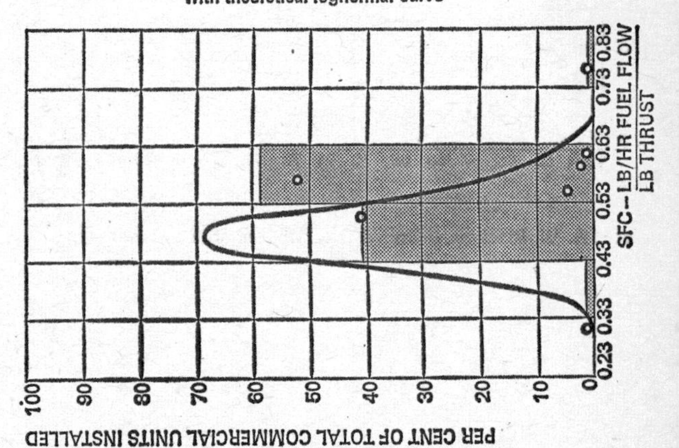

Source: A. W. Blackman, United Aircraft Research laboratories

Figure 14. *Variation of Specific Fuel-consumption Shape Parameters with Time*

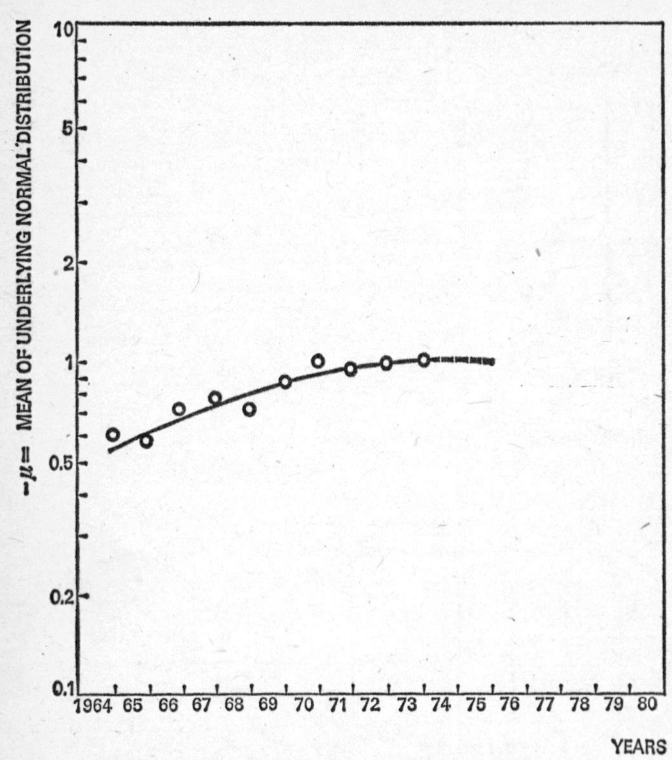

Source: A. W. Blackman, United Aircraft Research Laboratories

Figure 15. *Variation of Specific Weight Shape Parameters with Time*

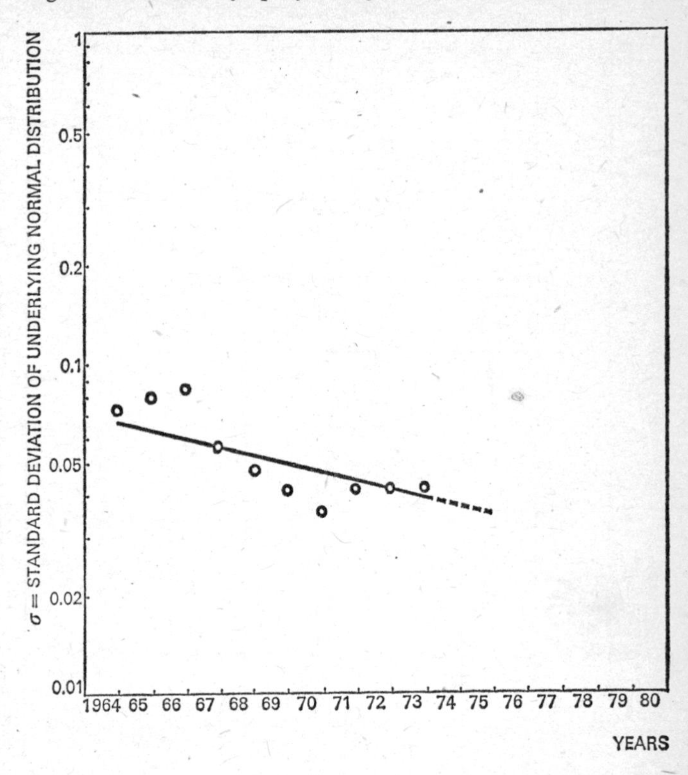

Source: A. W. Blackman, United Aircraft Research Laboratories

Figure 16. *Forecast of Distribution of Jet-Engine Specific Fuel Consumption in Commercial Market of 1975*

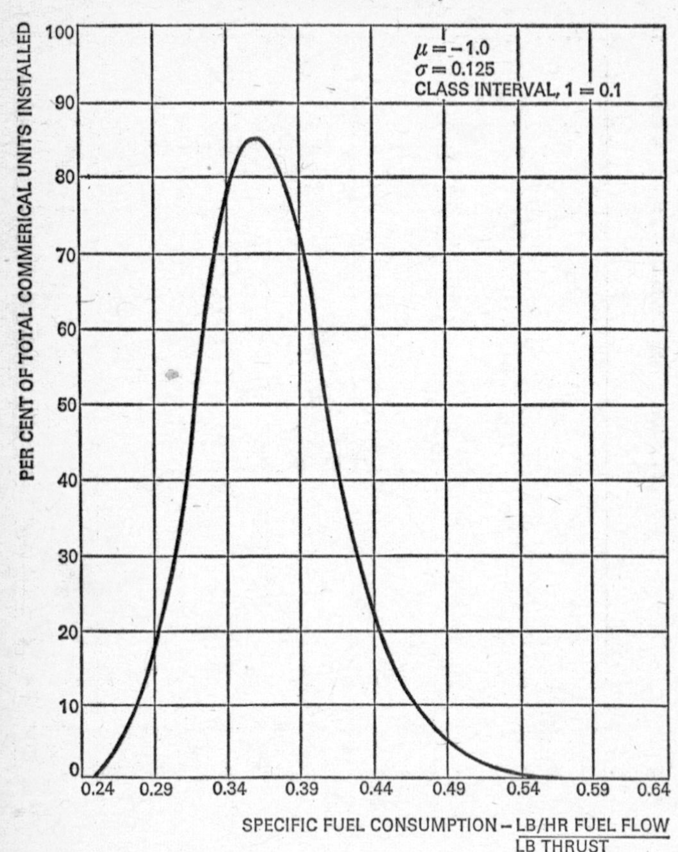

Source: A. W. Blackman, United Aircraft Research Laboratories

Figure 17. *Forecast of Distribution of Jet-Engine Specific Weight in Commercial Market of 1975*

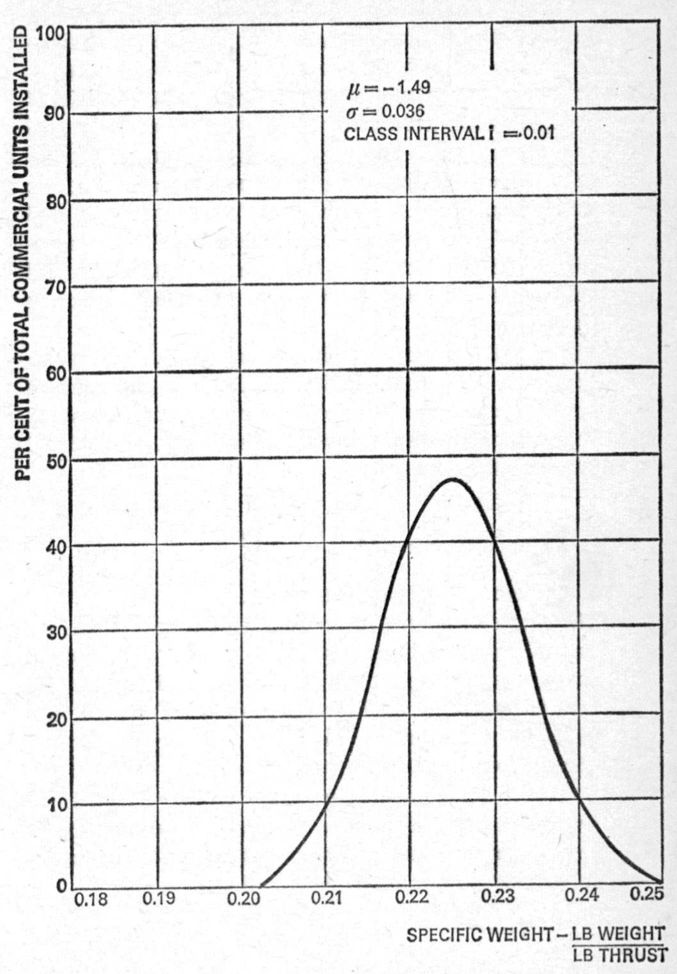

Source: A. W. Blackman, United Aircraft Research Laboratories

(ii) Segment those total market sales by the critical technological performance characteristics. Two were determined, viz. specific fuel consumption and specific weight. Plot the histograms of the frequency distribution of sales as functions of these two critical engine performance characteristics, for a series of selected years. For each of these series of histograms, cumulative frequency plots were constructed on lognormal probability paper. Figures 12 and 13 demonstrate stage (ii) for just one time interval in each of the two series calculated. Lognormal probability paper was used since the lognormal distribution seemed to fit the histograms and plots on such paper facilitate the graphical determination of the distribution parameters, which is stage (iii).

(iii) The cumulative frequency plot on such paper, as in Figures 12 and 13, yields a straight line, and the mean and standard deviation can be derived from the points where the fitted straight line crosses the probability scale values of 0·5 and 0·84, as follows:

$$\text{mean } (\mu) = \ln X_{0 \cdot 5}$$

$$\text{Standard deviation } (\sigma) = \ln\left(\frac{X_{0 \cdot 84}}{X_{0 \cdot 5}}\right)$$

where X is the parameter value.

(iv) The values of the means and the standard deviations for each distribution in each series on critical technological characteristics were plotted against time and extrapolated into the future. Figures 14 and 15 show two of the four plots made, for variations of μ for specific fuel consumption and of σ for specific weight.

(v) Using the extrapolated values for μ and σ obtained at step (iv), frequency distributions of sales as a function of the two critical performance characteristics were calculated for future years of interest by the equation:

$$Y = \frac{0 \cdot 398 \text{Ni}}{X\sigma} \exp\left(\frac{-(\ln X - \mu)^2}{2\sigma^2}\right)$$

where Y = value or ordinate

X = value of abscissa
N = number of items in sample (total units sold per year)
i = class interval
μ = mean of underlying normal distribution.

The final output for 1975, on each of the critical technological characteristics, is shown in Figures 16 and 17.

The introduction, through normex procedures, of the sales characteristics of the market as they relate to technological capabilities is of particular value in interpreting and even modifying direct extrapolations. Market-based data relationships will encompass many of the trade-offs which extrapolative approaches cannot handle directly. As for drawbacks – normex forecasting has the weakness that any basic extrapolation has in that it assumes historical determinants of trends will also be operative in the future. Furthermore, the technique is operable only where a strong relationship between sales and technology-performance characteristic exists. This currently restricts it to technology-intensive products, but refinements can perhaps modify this; there is no reason why the general methodology cannot be applied to marketing or advertising intensive products provided that the correct parameters are identified. Despite these shortcomings, it does furnish an effective measure of the forecast uncertainty by the computation of σ, and this is a considerable advantage in marketing as well as R & D planning.

CHAPTER 2.4

The Delphi Method of Time-scaling

THE Oracle at Delphi was a vital source of forecasts for the future, so it is not surprising that it should find a contemporary echo in one of the more charming and enjoyable techniques of TF. Although the Oracle's forecasts were merchandized with vestal virgins, they did little more than predict a particular course of events, and virtually no interrogation of the forecasters was permitted.

The Delphi method originated in a formalized manner in the work of Olaf Helmer in studies for the Rand Corporation. It was first reported on in 1963. Since then it has blossomed as a method for determining the pattern of expert opinion as to the likely timing of future events. It frequently combines within its procedures the initial development of scenarios, which we have already discussed in Chapter 2.3; so we shall exclude this element here. We shall examine the Delphi method for time-scaling those scenarios which have already been discerned, and for time-scaling other normative forecasts or morphological analytical outputs as desired.

Helmer's original Delphi forecast was concerned with six broad fields – scientific breakthroughs, population growth, automation, space progress, probability and prevention of war, and future weapon systems. Three rounds of forecasting were undertaken by six panels of 82 experts in all, as follows:

Round (i): Panel members were asked, by letter, to situate the 50–50 probability of realization of each scenario within one of a number of time periods into which the next fifty years had been divided. Answers of 'more than fifty years' and 'never' were also permitted.

The answers were charted to demonstrate the median forecast, and the various quartiles of the replies. The lowest and highest 25 per cent of forecasts represent the lower and upper quartiles respectively. Reasonable consensus was found for some 20 per cent of all scenarios (there were fifty) in Round (i).

Round (ii): Panel members were informed by letter of the outcome of Round (i), and in particular of the consensus. Dissenters were invited to explain the reason for their divergence and/or to modify their forecast. A selected group of the dissenting scenarios (some 80 per cent) was resubmitted and the rationale for each dissenting view was sought in light of the dissensus which had emerged at Round (i). Many forecasts moved closer to a consensus.

Round (iii): Panel members were again informed by letter of the outcome of the preceding Round (ii), and invited to proceed in the same way in the light of the newly revised forecasts. Some thirty-one or 60 per cent of scenarios managed to achieve a reasonable degree of time scale consensus by the end of Round (iii).

Some valuable information about the nature of the forecasts emerged. The quartile range of a forecast, i.e. the precision of the consensus, was approximately equal to the expected distance in the future as expressed by the median. Hence, for a median of x years hence, the lower quartile $\simeq 2/3x$ and the upper quartile $\simeq 5/3x$. This gives, for example, when a median forecast of the year 2000 is made in 1970, the lower quartile

$$\simeq 2/3.\,(2000-1970)+1970 \simeq 1990,$$

and the upper quartile

$$\simeq 5/3.\,(2000-1970)+1970 \simeq 1220$$

The quartile range was found to decrease with successive rounds of iteration, with an average ratio between Rounds (i) and (ii) of 5/8, i.e. consensus increased.

It may well be asked why such an elaborate method of getting estimates for the timing of future events is necessary. Surely we should simply ask people once, rather than harassing them into modifying their views through successive stages of a questionnaire procedure? The arguments advanced in favour of a

Delphi-type approach, however, are convincing.

The expert has built-in problems which are an occupational hazard, the direct result of his expertise. He will have an exceptionally good feel for historical trends in his own special field. He is only too well aware of the parameters which limit his progress in given directions. He is fully conversant with research work going on in his field and of the likely benefits to emerge therefrom. But he undoubtedly sees the world mainly through the eyes of his own speciality. Specialisms normally act as contributing elements to future directions rather than as their sole determinants and such specialist brilliance will often blinker appreciation of other contributing areas and even of their impact on his own technology.

When we add to this quite natural myopia a general human reluctance to predict development outside the specialist field for fear of ridicule, (or, more kindly, from a very real scientific humility) the views of specialists can be seen as suspect. A classic of specialist myopia was demonstrated by Lord Cherwell, Winston Churchill's scientific adviser during the Second World War when he insisted that the V2 rocket, which did not employ solid fuel propellants, would never fly. Lord Cherwell (then F. W. Lindermann) was an ardent advocate of solid-fuel rocket propellants, and stuck to his view despite photographic evidence to the contrary.

A committee approach, with face-to-face discussion about the timing of forecast events, can go some way towards avoiding the myopia of the brilliant specialist thinking alone. None the less, it also has drawbacks. New specialists add further parameters which constrain progress in any given direction. Furthermore, if the committee is composed of members of differing status they may be reluctant to expose themselves to ridicule. Junior members, unless precocious, will normally be unwilling to chance their arm in the presence of their superiors, even when the most effective applications of group dynamics and organic leadership are made. Finally, the bandwagon effect and the influence of the glib, fast-talking committee members can upset any valid movement to real consensus. Even more, committees can give rise to a consensus as a result of face-to-face compromise when dissensus would more accurately describe the true state of expectations.

The Delphi method can draw on that which is valuable in the committee approach, in particular the benefit of perceiving the line of reasoning of others, whilst avoiding the social compromise which normally occurs at a meeting. Individual views can be argued, abandoned or sustained behind the mask of anonymity, without loss of face and without being affected by the personality, reputation or seniority of the protagonists. The method also permits the individuals participating in the panel to consult with their colleagues (not fellow panel members), should this be deemed desirable.

As thus ennunciated the Delphi method is attractive; but it has several important weaknesses. The opinion that consensus is not forced by social pressure on the divergent when face-to-face confrontation is avoided is not wholly convincing. Less convincing still is the aggregation of all the views of all panel members as though they are all of equal value. Few people would deny that some panel members' views are undoubtedly more valuable in certain sectors than others no matter how carefully the panel is recruited. Few rules can be given for the selection of experts save that the judgement is probably most sensibly exercised by a small group rather than by one individual.

Stratification can, of course, be introduced into the panel to ensure that only certain parts of the study go to seemingly qualified members. This was how Helmer's six original areas were forecast. But this method tends to conflict with arguments cited above against the employment of the contemporary field specialist. However, this might perhaps be mitigated by the weighting of views expressed.

The weighting of opinions within a panel has now been undertaken on several occasions within a consistent pattern. Panel members have been invited to rate their own competence in each sector to which they respond and, if they feel totally unequipped, to make a nil return in a particular area of the study. Helmer has used a system with competence rankings from 1 to 7 over a 7-question sequence. The panel members were enjoined to ensure that they ordinally ranked their competence to answer each of the seven questions – i.e., that they used each competence score 1 to 7 once only. This sounds less satisfactory than a procedure which offers, say, a competence range of 0 to 10, and allows panel members to indicate their self-perceived competence on each question in isolation from the overall sequence.

A weighted forecast can be achieved by replicating forecasts in proportion to the forecaster's competence factor prior to analysis. Alternatively, results can be displayed separately for various levels of self-rated competence. In two recent studies of economic forecasting it was found that the views of self-rated experts were substantially better than the views of the total panel over 65 per cent of the time, i.e., their median forecast was closer to the true outcome when it emerged.

Figure 1 demonstrates how this can be accomplished on a Round 1 questionnaire. The questionnaire also demonstrates an important extension of response measurement. Panel members are not asked to simply make a single point forecast of the 50–50 probability of achievement of a given scenario. They are asked to ascribe forecasts for different levels of probability of achievement – say 0·1, 0·5, 0·9.

Our panel member's hypothetical replies demonstrate that he feels himself especially competent in the plastics field, less so in computing applications and very much a layman in the field of mass retailing. The second scenario on retailing does, in fact, demonstrate a very important weakness of the Delphi method, which the competence scaling does not resolve completely: namely, the respondent can be without relevant factual information in the light of which his competence could dramatically increase. If one firm rule can be made about what *not* to include in a Delphi-type questioning sequence it is almost certainly that forecasts which are based on relatively easily determined facts are better not included, unless the facts are made available. In scenario 2, on retailing, I am certain that most panel members would want to know the *existing* percentage at the time of making their forecast.

Figure 1. *Delphi Questionnaire with Competence Self-rating and a Range of Probabilistic Forcasts*

Self-Rating of Competence in the area 0–10 (max.)	*Scenarios (from 1970)*	*Forecast Date for Various Probabilities of occurence*		
		0·1	0·5	0·9
3	Over 50 per cent of all consumer spending made in large localized shopping centres by . . .	1980	1985	1990
1	By what year will large localized shopping centres offer 95 per cent of all categories of consumer goods?	1990	1995	2000
6	Automated highways and adaptive car autopilots for 5 per cent of all travel in the UK by . . .	1985	2000	2020
8	Moulded plastic garments meeting needs of 5 per cent of UK market by . . .	1986	1992	1999
4	Personal identification reliability for banking to 99·99 per cent by . . .	1979	1984	1996

THE PHRASING OF SCENARIOS

We have already indicated in Chapter 2.3 the procedures for generating glimpses of futures. In the time-scaling content of the Delphi method, the precise formulation of questions becomes critical. Each and every panel member must understand the meaning of each question in an identical way. This is not to say he will not bring differing assumptions to bear in making a forecast, but that the stated future must be identically perceived. This is no greater a problem than is traditionally encountered in the social sciences or particularly in marketing research. The best advice I can offer to questionnaire writers is to read Sidney Payne's brilliant book *The Art of Asking Questions*. He reviews the pros and cons of open ended and forced questioning procedures, alternative phrasings and the 'are you still beating your wife?' variety of leading questions. He explores the virtues and problems of brevity and revels in the search for the *mot juste*; he concludes with a list of 100 considerations.

Parker, of the Hercules Powder Company in the UK, who recently completed a major Delphi study in cooperation with firms throughout the industrial chemical industry, has given an explicit description of his problems in formulating questions. He has also focused attention on one of the great advantages of the iterative questioning procedures that are used. They afford the opportunity for panel members to feed back the necessary critique of questions in order to ensure that all are indeed forecasting the same future event. Provided only Round (iii) forecasts are utilized, any shortcomings can be overcome. Obviously this must not be taken too far. An abysmal Round (i) questionnaire will be dismissed as incompetent and the probe of the future will get no further.

The Hercules study was by no means incompetent in Round I or at any other stage. It was, in the view of the panel of thirty experts, a great success. Its essential starting elements were:

(i) *an initiator* (E. T. Parker);
(ii) *an adequate panel* from the entire industry – sixty were invited and 50 per cent agreed to participate;
(iii) *a clearly defined objective*;
(iv) *an acceptable questionnaire*;
(v) *an analytical framework*;
(vi) *an agreed timetable* at the outset; and
(vii) 'the cheek of the devil and the foolhardiness of an amateur bull-fighter'.

Figure 2 reproduces the full Round (i) questionnaire sent out by Parker. It will be seen that competence ratings were not formally used but panel members were simply invited not to reply where they were on unfamiliar ground. Parker did, however, ask panel members to ascribe forecast at different levels of probability in the future but a specific set of three dates was imposed on respondents. This makes the study less desirable from a methodological viewpoint but certainly more manageable.

A considerable body of notes and advice for completion are included at the end of the questionnaire. This Round (i) questionnaire was the outcome of correspondence with panel members on an earlier draft. Even so, problems were to emerge for modification at a later date. Parker describes his experience thus:

Even at the preliminary stages, it was crystal clear that much depended on how the questionnaire questions were formulated. . . . I may be quite convinced that the petrochemicals industry is going to grow like a cuckoo in the nest and that production of organic chemicals by other routes is therefore doomed to dwindle. But if I asked questions like 'How much do you expect petrochemicals to grow' and 'How much do you expect other organics to decline' I was to some extent putting words into the other fellow's mouth.

Another important consideration was that of getting replies in a form convenient for reporting back. Quite clearly, replies in quantified form are desirable, because they can be analysed by standard mathematical techniques and fed back in concise tabulated form. But numbers, to be meaningful, must correlate back to a properly understood base. For this reason, in the case of the very broad questions in Section A, on well-established divisions of the industry, it was decided to make the reference base the index of production for the most recently-reported year, namely 1967.

When we come to innovations, as in Section B, there exists no such reference base, since by definition the current production index of a future innovation must be zero. In this case, it was judged most sensible to call for forecasts in terms of actual tonnage of future production, although there is a big weakness here – one can envisage a terribly important new material, perhaps like carbon fibres, which will be very costly indeed and sell in very small quantities by weight and these under this system would not show up in anything like their real significance . . .

Section C moved on to classification by end-use. After much cogitation, it seemed best here to quantify the effect of innovations in terms of the total percentage share they might capture of their defined markets. I may add that, acting on suggestions from panel members, this section was expanded after the first round, to rectify a glaring omission, 'Construction – alternative manufactured organics versus *wood*', and so subdivide agricultural chemicals into three categories.

All this was pretty down to earth, although you'll notice that we did put in, in all cases, space for conjecture about the nature and/or sources of the innovations foreseen. Then, to open the thing right out to the broadest possible speculation, we included Section D for written replies only which did not lend themselves to being quantified.

In addition, for the first round only, we inserted a section in which the respondents were invited to record their formal qualifications, their experience, and the extent to which they were occupied in long-range studies.

In mounting an exercise like this, it quickly becomes apparent that words are terribly important. Take the phrase 'manufactured organic chemicals', which sounds simple and straightforward enough. But what do we mean by 'manufactured'? Rosin and cellulose. for example, are widely regarded as industrial organic chemical raw materials, but they are obtained from natural sources, rather than manufactured in the strict sense. In inorganics, limestone is quarried and quarrying is extraction not manufacture. It was necessary to lay down a distinction between chemicals *processed out of* natural raw materials, which qualified for inclusion and chemicals which were merely *refined*, which did not, except under the minerals heading.

Under 'Outlets' you might suppose the word 'transport' was pretty unambiguous. But it was pointed out that great tonnages of liquid chemicals are being, and will be increasingly, transported via pipelines, and the question was asked, 'Was this transport within the meaning of the Act?'. The answer, in this case, was 'No', but this had to be made clear to all. Later in the exercise it was realized that some innovations might apply only to sub-divisions of the transport industry – for example, vacuum propulsion could only work in tunnels – and it was decided to split the heading into two sub-sections, 'constrained transport' and 'free-moving transport'. In like manner, in the construction section, a split was made into 'load-bearing' and 'non-load-bearing' sub-sections.

Not to put too fine a point on it, definitions proved to be at once the most critical and the most exacting element in the entire exercise.

Not only in the questionnaire, but also in handling the written replies, problems of terminology exist. Under 'new textiles' one respondent asserts that major characteristics will be 'better physical and aesthetic properties', another 'improved strength and drape' and a third 'improved feel and permanent set'. Do these mean the same thing? And if they do, which words does one choose to report back in summary? Likewise, one member expects synthetic food proteins to be produced by 'photo-synthesis', another by 'micro-organisms', a third by 'enzymes' – how far are these all saying much the same thing in different ways?

One answer would be to compile a comprehensive glossary of definitions and issue it to the panel before the exercise starts; but it would be so big as to be forbidding, and one can hardly expect volunteer panel members to spend hours with a book on terminology to make sure every word they use is on the approved list. At the same time, it is essential that some basic definitions be agreed in advance, or results may be meaningless. I would suggest that all technical terms used on the printed questionnaire should be defined, and panel members asked specifically to observe the definitions. For the rest, it seems to me that there is no substitute for good old common sense.

Source: Parker, E. F., (1969).

Figure 2 *Questionnaire for a 'Delphi' Exercise*

FIRST ROUND 'DELPHI' EXERCISE – FUTURE PATTERN OF UK INDUSTRIAL – CHEMICAL INDUSTRY*

Respondents are asked to enter their own considered opinions on the issues listed below.

Very important. The scope of the questionnaire is broad. It is suggested that respondents omit entries on issues where they hold no views. The value of the exercise would be undermined if respondents merely 'thought of a number' so as to complete an entry where they were on unfamiliar ground. However, considered views in areas outside a respondents' direct experience are valid for inclusion.

*Reprinted by permission of the Hercules Powder Company.

Enter your forecast production index (1967 = 100) for each year.

	1980	1985	1990
A1 All manufactured organic chemicals			
A2 Primary organic chemicals obtained from petroleum			
A3 Manufactured organics from non-petroleum sources			
A4 Manufactured inorganic chemicals (including minerals chemically refined)			

1000 ton units

	1980	1985	1990
B1 (a) Organic chemicals from any major source(s) other than petroleum (e.g. air/water)			
(b) New source(s) envisaged.... .... (c) Date(s) of first commercialization			
B2 (a) Novel inorganic chemicals from new fundamental technology(ies) (e.g. inorganic polymers)			
(b) New technology (ies) envisaged .... (c) Date(s) of first commercialization			

	per cent of total market served for each year		
	1980	1985	1990
C1 *Transport* (a) New sources of automotive energy (e.g. fuel cells) versus petrol and oil			
(b) New energy source(s) envisaged .... Date(s) of first commercialization			
C2 *Transport* (a) New mobile loadbearing devices (e.g. air cushions) versus rubber tyres			
(b) New device(s) envisaged			
C3 *Construction* alternative metals versus iron and steel			
C4 *Construction* alternative non-metals (e.g. plastics) versus iron and steel			
C5 *Construction* alternative organics (e.g. plastics) versus mineral products (bricks, concrete, slate etc.)			
C6 *Foodstuffs* (a) synthetic edible proteins versus natural			
(b) Synthetic protein source(s)			

	1980	1985	1990
C7 *Foodstuffs* (a) synthetic nutritive carbohydrates versus natural			
(b) Synthetic carbohydrate source(s)			
C8 *Foodstuffs* (a) synthetic edible oils versus natural			
(b) Synthetic oil source(s)			
C9 *Textiles* (a) man-made fibres of kinds not so far commercialized			
(b) General nature of the new fibres			
C10 *Agricultural chemicals* (a) new ag. chemicals not so far commercialized			
(b) Sources of new chemicals .... (c) Purposes for which developed			

Section D
Broad Speculation

D1 Advances of broad chemical interest not included above, with anticipated dates of commercialization ......
....
....
....

D2 (a) Will fresh fundamental chemical principles capable of commercialization be discovered between now and 1990? YES/NO

(b) In what fields of study, if any, do you think such discoveries will be made?
....
....
(c) And when ....

Section E. Personal

E1 Current duties include (a) long-range planning/ technological forecasting
(b) full-time/part-time/not at all.

E2 Technical qualifications are in pure chemistry/applied chemistry/chemical engineering/ mathematics/ statistics/ economics
Other ....

'DELPHI-TYPE' EXERCISE ON UK CHEMICAL INDUSTRY (HERCULES)

Notes

In answering numerical questions, please give *single figures* only in the appropriate boxes, not ranges. The figure should be your own 'most probable' judgement at the time of responding. (Reports will show the range embraced by the replies from all the panel, and thus indicate any areas of wide disagreement.) It is expected that your quantified replies may well be amended during rounds 2 and 3 in the light of fresh thinking stimulated by the intervening panel reports fed back to you.

It will be assumed throughout that your quantified predictions take account of the technological developments you consider most likely, including those which come to your notice during the course of the exercise.

Answers to numerical questions should relate to UK industry only (but take account of new developments elsewhere in the world).

Answers to 'speculation' questions should relate to new developments regardless of where in the world you think they may start.

Definitions

Manufactured organic chemicals. Include all organics, regardless of raw material source, which differ in a material chemical sense from the raw material. Commercially pure chemicals fractionated from natural chemical mixtures qualify, but refined raw materials, still substantially the same mixtures, do not.

Transport. Transport which is itself mobile qualifies. Conveyance via static structures, e.g. pipe lines, does not.

Proteins. Include all proteins made by synthetic industry. Allow in 'total market' for proteins recovered from natural sources not commercially exploited at present. Identification of source will help.

Petroleum. Includes other current natural sources of organics, e.g. shale oil and natural gas, but excludes coal.

Manufactured organics from non-petroleum sources. Sources include established raw materials such as molasses, cellulose, coal and coal tar, rosin and turpentine, vegetable matter generally.

Construction. Is meant to include all housing, office, factory, bridges, tunnels and the like, but excludes roadmaking, process plant, fabrication, transportation, machinery, etc. If mobile housing is expected to become significant, it should be included.

'Synthetic' edible proteins. Covers proteins for human consumption, from any purely synthetic process or by enzyme/bacteriological/biological processes or others, applied to non-protein starting materials. It excludes proteins 'won' by new methods from protein-bearing sources not exploited today.

Textiles. Includes all synthetic fibres plus any non-woven synthetic fabrics, films, etc., used as alternatives to traditional textiles.

Readers might care to grapple with a few questions in the draft Delphi questionnaire shown in Figure 3, for motor vehicles. This was developed for training purposes by the author in association with H. Jones and J. Grigor. It seems to show plenty of questioning errors in an area where more of us consider ourselves expert than almost any other. All the words set in italics seem to be troublesome.

Figure 3. *Training Draft of a Badly Phrased Delphi Questionnaire*

N.B. the motor car is the private motor car and questions relate to *Britain* only.
Please rate yourself for competence in answering each question *between 0 and 10* in the circle provided.
Please circle *answer preferred.*

1. Will the number of licences issued for *vehicles* on the road be *limited* in the next:

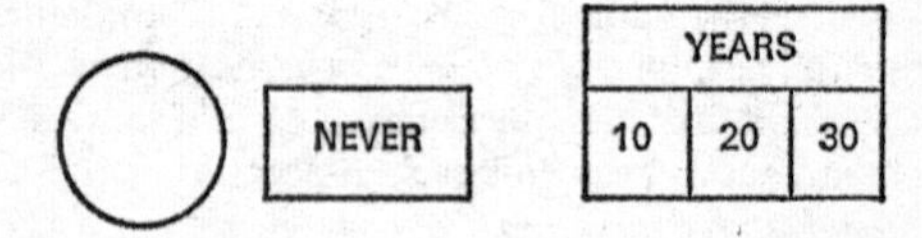

2. For any reason do you see the *average size* of *vehicles* on the road decreasing in the next:

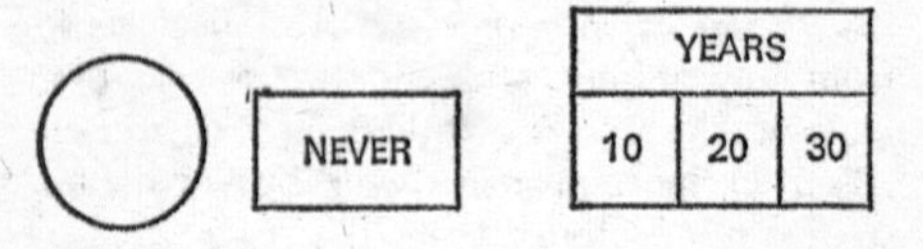

3. Do you expect safety regulations which would make illegal *vehicles* not capable of *conversion to satisfy statutory requirements* in the next:

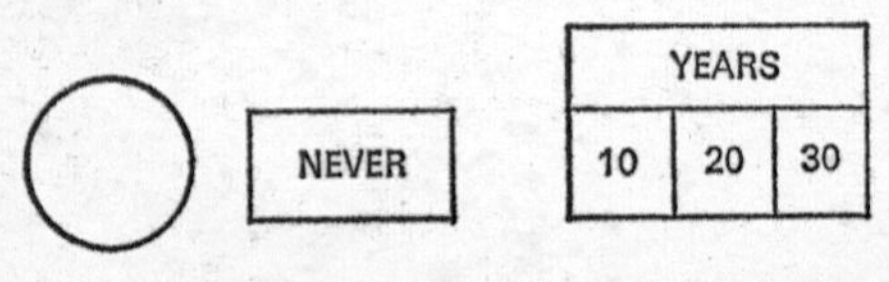

4. *If Yes to Question 3*: What *period of grace* would you expect for conversion?

YEARS					
0	1	2	3	4	5

5. Will *air-pollution regulations* for *vehicle* exhausts come into force in the next:

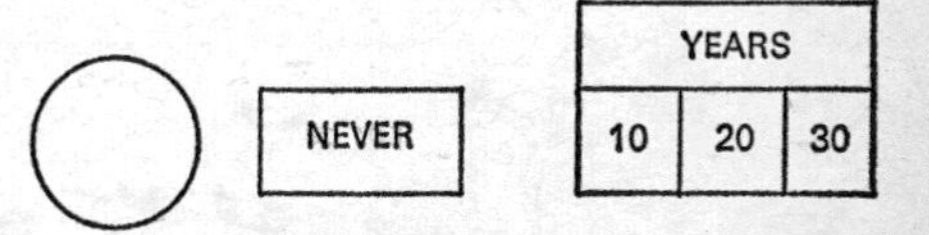

6. Do you *expect* private cars to be banned from *city centres* in the next:

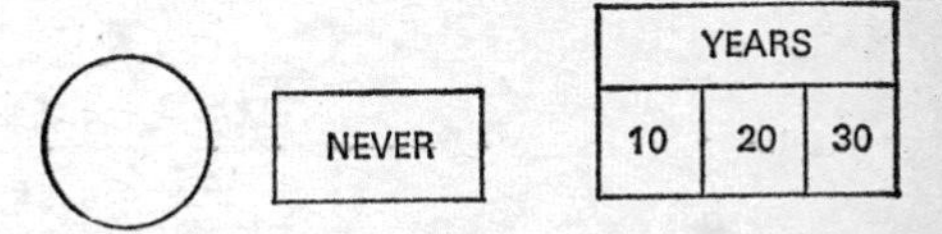

In Questions 7, 8 and 9: When completing boxes, give answer for each year (10, 20, 30)

7. In view of your answers to the foregoing, and *considering the car itself*, what percentage of *motor vehicles* will be propelled by:

	YEARS HENCE		
	10	20	30
	per cent	per cent	per cent
Internal combustion			
Steam			
Gas Turbine			
Electricity			

8. In what percentage of *vehicles* will the wheel be replaced by other methods of *movement*?

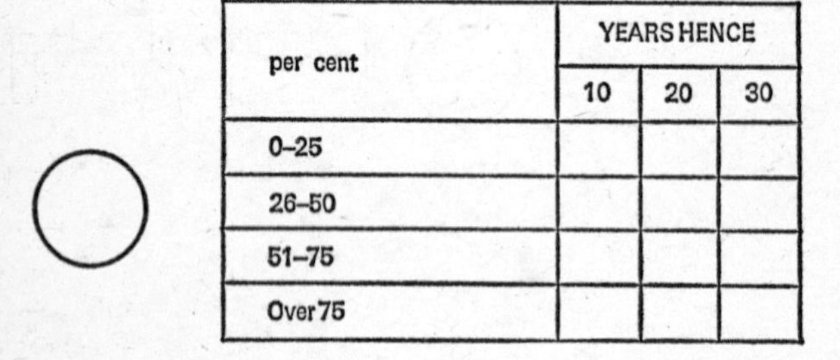

per cent	YEARS HENCE		
	10	20	30
0–25			
26–50			
51–75			
Over 75			

9. In *what percentage* will metals be replaced by plastics, in Body?

per cent	YEARS HENCE		
	10	20	30
0–25			
26–50			
51–75			
Over 75			

in engine, transmission, etc.?

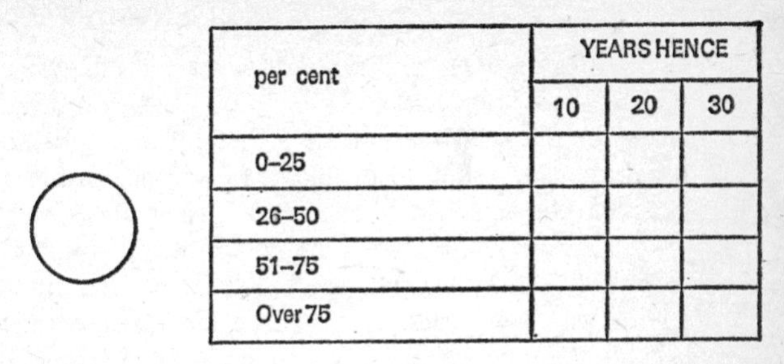

per cent	YEARS HENCE		
	10	20	30
0–25			
26–50			
51–75			
Over 75			

10. Do you foresee the *advent* of a car *in popular use* which requires no maintenance other than charging with fuel and replacing worn parts in the next:

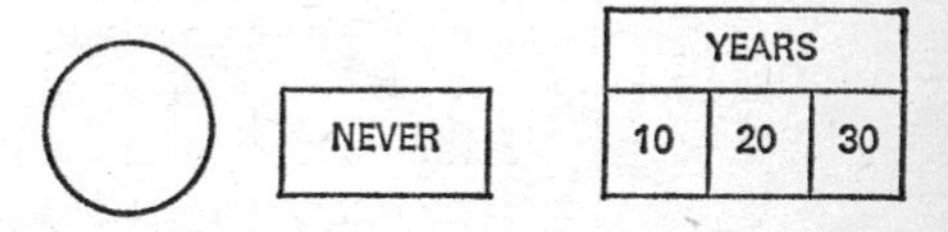

11. Do you foresee the *advent* of permanent sealed lubrication in engines in the next:

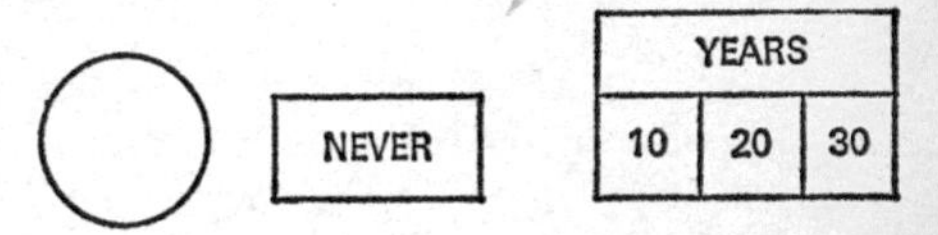

12. Do you foresee the *advent* of the car *in popular use* which never requires external cleaning, in the next:

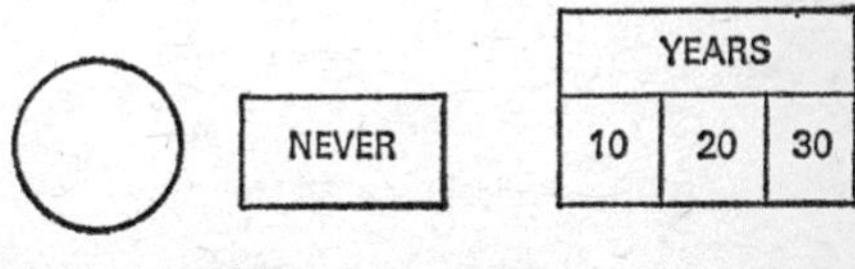

MOVEMENT TOWARDS CONSENSUS AND OUTCOME

Movement towards a Delphi consensus is achieved by feeding back the detailed logic and reasoning of dissenters from the general opinion at each round of the process of iteration. Coping with this presents substantial problems of editing. In the main, however, it can be achieved in two ways. Either divergent panel members are contacted immediately following receipt and analysis of their questionnaires and invited to write briefly in support of their view which can then be incorporated into the following round, or alternatively, explanations for divergence can be elicited during the course of the next round and fed back to others in the following stage. The former method is to be preferred in general but can involve respondents in not just three rounds but two intermediate explications in addition.

Outcome display of Delphi findings is especially absorbing since its originators initiated a new approach to data presentation, a horizontal histogram or bar-chart approach. This is often accompanied by a verbal description of the movement between rounds. Parker, reporting on Round III of the Hercules Powder Study for example, although he decided not to employ the novel data presentation procedure, commented and displayed his outcome for Questions C6, C7 and C8 as shown in Figure 4 (a). Figure 4 (b) shows an intermediate feedback stage for a study at LTV Inc.

The novel approach to outcome display is indicated in the series of final round forecasts which are given as Figure 5, 6, 7 and 8. Figure 5, from Ozbekhan, examines technological progress in automation as predicted by a Delphi panel. The upper and lower quartiles of each range of forecasts, which were single point, have been removed. Hence the bar indicates just the middle half of the answer given and the peak value represents the median forecast.

Figure 4(a). *Hercules Delphi outcome for Foodstuffs: Round III*

		Per cent of total defined market served for each year							
		neglig-ible.	1–4	5–9	10–14	15–19	20–24	25–29	Others
C6. 'Synthetic' protein	1980	6	7	1	0	0	0	0	0
	1985	1	7	5	1	0	0	0	0
	1990	1	4	3	4	2	0	0	0
C7. 'Synthetic' nutritive carbohydrates		'Not in prospect' – no dissentients							
C8. 'Synthetic' edible oils	1980	12	0	0	0	0	0	0	0
	1985	7	4	1	0	0	0	0	0
	1990	2	9	0	0	0	1	0	0

Reproduced by permission of the Hercules Powder Company

There was a slight hardening of opinion since Round II as regards the long-term market share for 'synthetic' proteins as defined, and a notable decline in long-term market percentage rating for 'synthetic' edible oils.

Most probable sources of 'synthetic' proteins were cited as follows:

(i) Directly or indirectly from petroleum and/or natural gas – fifteen positive replies. This includes two who expect the feedstock to be petrochemicals rather than raw petroleum (probabilities 8, 5, dates 1980, 1978 respectively). Another member specifies microbial processes (a) on petroleum hydrocarbons (8, 1975+) and (b) on natural gas (3, 1980) – treated as two entries. One respondent ascribes probability 4, but qualifies 'unlikely'. A further member predicts use of n-alkanes (1978), methane (1980) and naphtha (1985) all probability 9.

Overall average probability rating 7; average year of UK commercialization 1976.

(ii) Three members cite as sources vegetation, including the development of special vegetable growths. Two assign probability 8, year

1980; the third 2 and 1975, mentioning 'low commercial barriers' as compared with oil and natural gas sources.

(iii) Two respondents envisage marine sources, including fish, by 1980, with probability ratings 8 and 9.

(iv) Micro-organisms are cited by two members (excluding the entry citing microbial action on petroleum, etc., already mentioned), one working on oil residues base, 'available via animal feeding' (8, 1975) and the other on cellulose (5, 1985).

(v) Another respondent cites cellulose waste as source, route unspecified (2, 1980s) qualifying with 'very limited acceptance'.

(vi) One other respondent cites photo-synthesis (8, 1980).

Most probable sources of 'synthetic' edible oils commanded very few entries. Four respondents cited petroleum and/or natural gas with probability ratings 7–9, placing the year of commercialization most probably in the late 1980s. Of these, one asserts that the source will be important by 1990 (predicting market share 20 per cent). Another, who points to the rising costs of natural products such as whale oil, specifies microbiological action on selected hydro-carbon feedstocks, with probability ratings 3, 6 and 9 for 1980, 1985 and 1990 respectively.

Marine sources, fish and seaweed have one mention (7, undated); cellulose sources have two, one specifying enzyme processes (2, 1980), the other unqualified (1, 1990); enzyme action on carbon dioxide is cited once without ratings. One member mentions soya as a source, 'already developed'.

Source: Private communications from Mr Parker to the author in September 1969

Figure 4(b). *Intermediate Feedback of Results in a Delphi Exercise*

Scenario; A majority of new home construction will include internal power (that is, fuel cells, solar cells, etc.) by the year:

ARGUMENTS PREVIOUSLY OFFERED: (Never) Power supplied from nuclear plants will be too economical. (Never) It will not be economically feasible to produce a unit of this type and be in competition with a central powerplant. (Never) Only probable possiblity is a safe, reliable, compact, cheap nuclear powerplant. Safe reliable, and compact it may be, but cheap it won't be. (2010) Community power sources will be more economical and cheaper to the user until a meaningful breakthrough is made. (2006) I agree that the only possibility is a reliable nuclear power unit in the home. This can't occur before the year 2000. Also consider the efficiency and cheapness of centrally generated and distributed power that is being used now. (2000) Assuming that this question refers only to the USA. (2000) If internal power assumes no distribution lines of any kind. (1990) Experiments have been made on the concept of this project for over seven years. Most of the gross concept has been worked out, only the details remain for solution. (1990) I think the non-thinkers who say 'never' are ridiculous, for it is just a matter of time and money. Drawing power from long distance with lines running everywhere is NOT in the future with our rapidly growing power needs. (1975) The American Gas Association has conducted successful experiments with natural gas fuel cells and contends that they will be competitive with conventionally generated and distributed electricity by 1970 or 1971. See Science-News November 1968.

Median; 1998 *Inner quartile;* 1995–2000 *Extremes;* 1975 & 2010 *Per cent Nevers;* 9·1 Your new estimate:

Your arguments:

Source: McLoughlin, W. G., (1969), 'Product Cycle Planning at Ling-Temco-Vaught Inc.', *TF Conference*, Lake Placid Club, September.

Figure 5. *Ozebekan's Delphi Timescale for Progress in Automation*

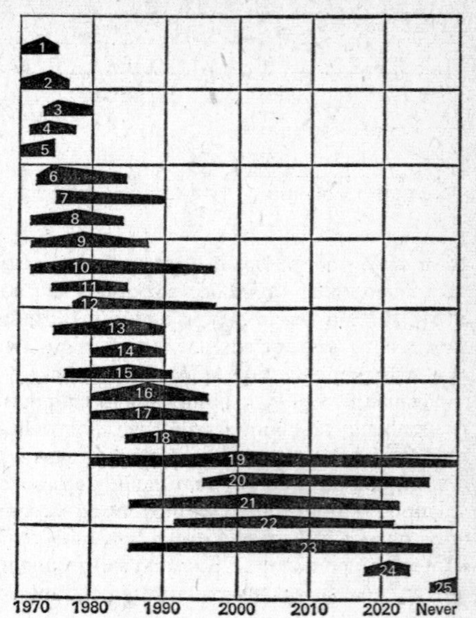

1 Increase by a factor of ten in capital investment in computers for automated process control.
2 Air traffic control – positive and predictive track on all aircraft.
3 Direct link from stores to banks to check credit and to record transactions.
4 Widespread use of simple teaching machines.
5 Automation of office work and services, leading to displacement of 25 per cent of current work force.
6 Education becoming a respectable leisure pastime.
7 Widespread use of sophisticated teaching machines.
8 Automatic libraries looking up and reproducing copy.
9 Automated looking up of legal information.
10 Automatic language translator – correct grammar.
11 Automated rapid transit.
12 Widespread use of automatic decision making at management level for planning.
13 Electronic prosthesis (radar for the blind, servomechanical limbs).
14 Automated interpretation of medical symptoms.
15 Construction on a production line of computers with motivation by 'education'.
16 Widespread use of robot services.
17 Widespread use of computers in tax collection.
18 Availability of a machine which 'comprehends' standard IQ tests and scores above 150.
19 Evolution of a universal language from automated communication.
20 Automated voting, in the sense of legislating through automated plebiscite.
21 Automated highways and adaptive automobile autopilots.
22 Remote facsimile newspapers and magazines printed at home.
23 Direct electro-mechanical interaction between man and computer.
24 International agreements which guarantee certain economic minima to the world's population as a result of high production from automation.
25 Centralized (possibly random) wire tapping.
(Source: Ozbekhan, *Science Journal*, October 1967).

Figures 6 and 7 describe two Delphi studies undertaken in Britain by ICL and reported by Hall in 1969. He has not truncated his bars, but the upper and lower quartiles were removed from replies and the peak value again represents the median forecast. Single points were again used. On balance, it is probably preferable to truncate the bar since it acts as a reminder that upper and lower quartiles have been removed.

Figure 8 displays in another way the outcome of Delphi studies at TRW, the Thompson-Ramo-Wooldridge Corporation, which were reported by North and Pyke.

Figures 9 and 10 go behind the charting procedure to indicate the TRW method of analysis which was undertaken by computer. For Event 201030, the scenario that *electric cars using fuel-cell power or fuel-cell/battery combination will be marketed commercially*, we can see the full gamut of questionnaire procedures in use. Familarity is here used as synonymous with our earlier concept of competence with a 3-point weighting. Assessments of social desirability as well as technical feasibility are also obtained and a split of probabilities – first, that the scenario will ever materialize; and then the time-scales for 0·1, 0·5 and 0·9 probabilities, if we assume that the event will occur.

Figure 9 is one panel member's response, member 304. Figure 10 shows the feed-back data for each panel chairman and/or member to utilize in the next round. It displays all the detail fo his own panel and summarizes for all other contributing panels for the scenario in question. Panel member 106, who has been given an asterisk, seems sufficiently out of line for the originator to seek a logical explanation of his forecast to attempt to understand its divergence.

Figure 6. *Hall's ICL Delphi Time-scale for Utilization of Computers*

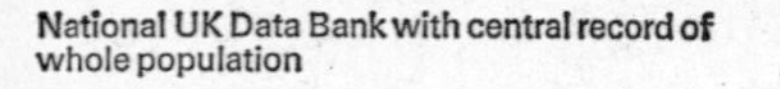
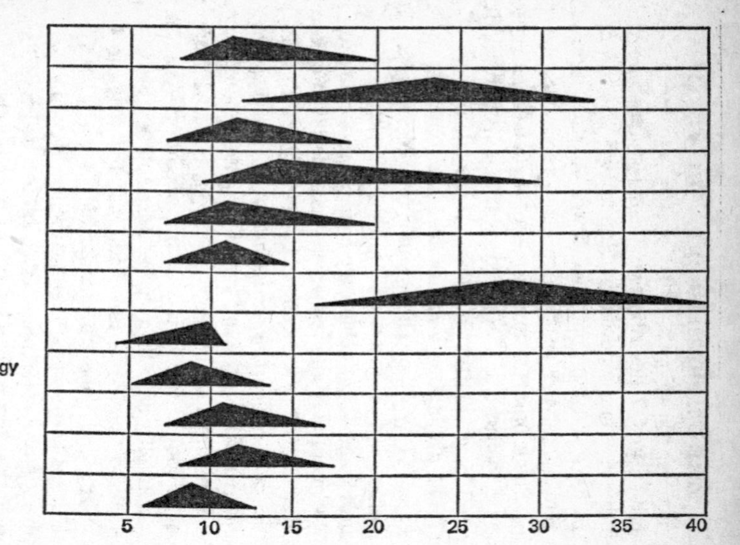

Source: Hall, P. D., in Wills, G.S.C. et al., (1969), p. 200.

Figure 7. *Hall's ICL Delphi Time-scale for Development of Computer Technology*

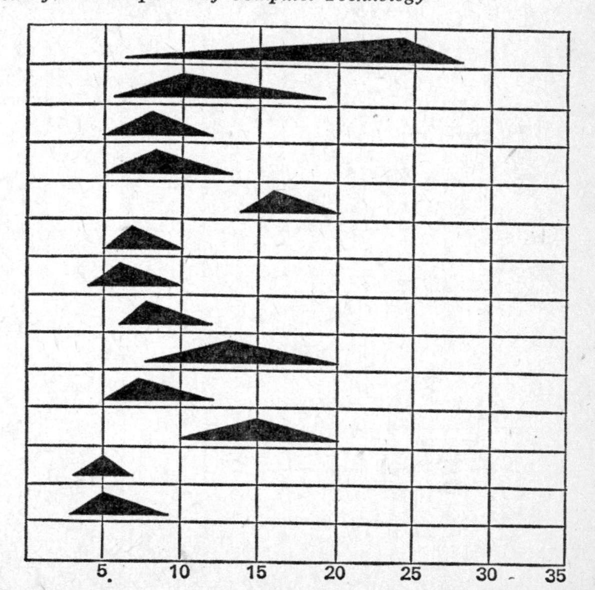

Source: Hall, P. D., in Wills, G.S.C. et al., (1969), p.201

Figure 8. *Some Selected TRW Delphi Time-scales*

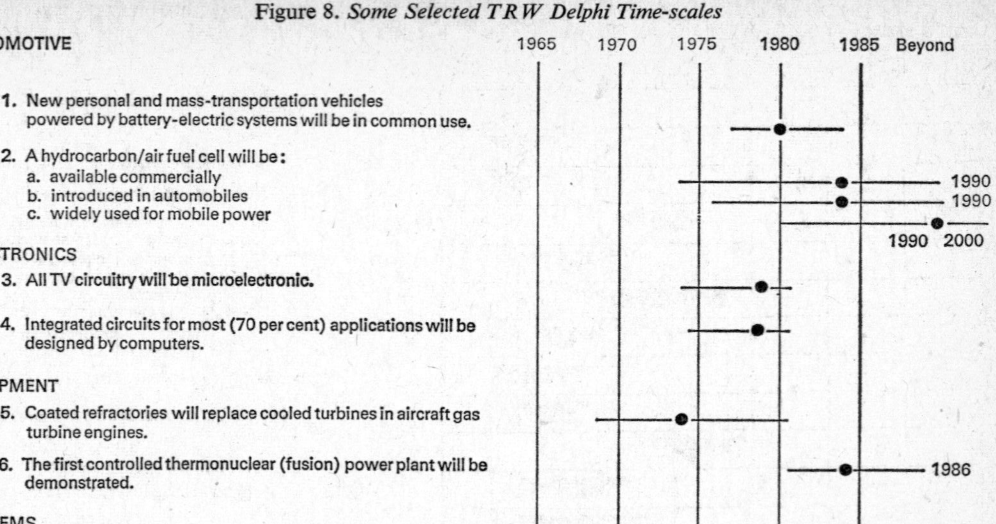
AUTOMOTIVE
1965
1970
1975
1980
1985
Beyond
1. New personal and mass-transportation vehicles powered by battery-electric systems will be in common use.
2. A hydrocarbon/air fuel cell will be:
a. available commercially
b. introduced in automobiles
c. widely used for mobile power
1990
1990
1990
2000
ELECTRONICS
3. All TV circuitry will be microelectronic.
4. Integrated circuits for most (70 per cent) applications will be designed by computers.
EQUIPMENT
5. Coated refractories will replace cooled turbines in aircraft gas turbine engines.
6. The first controlled thermonuclear (fusion) power plant will be demonstrated.
1986
SYSTEMS
7. Large-thrust, air-augmented manned recoverable booster rockets (aerospace planes).
8. Weapons utilizing laser technology will be in use.
1995

Source: North, A. Q. and Pyke, D. L., (1968)

Figure 9. *Individual Panel Member's Forecast in TRW Delphi Method*

Probe Category 9. Transportation			*Panel Member 304*				
Event Number	Event Description	Panels Evaluating	Familiarity	Desirability	Feasibility	Probability of Event	Dates
201030		01 02 04 09					
Electric Automobiles using Fuel-cell			1 Fair	1 Needed	1 Simple		·1 Date 1979
Power or Fuel Cell/Battery Combination			2 Good	2 Desirable	2 Possible	0·7	·5 Date 1985
will be Marketed Commercially			3 Excellent	3 Undesirable	3 Unlikely		·9 Date 1990

Source: North, H. Q., and Pyke, D. L., (1968)

Figute 10. *Evaluation of Scenario by all Panel Members in TRW Delphi Method*

ELECTRIC AUTOMOBILES USING FUEL CELL OR FUEL CELL/BATTERY COMBINATION WILL BE MARKETED COMMERCIALLY

Panelist	*Familiarity*	*Desirability*	*Feasibility*	*Probability of Occurence*	ALREADY	*Probability Dates Assuming Event*							*2000 and Beyond*		
						1970	1975	1980	1985	1990	1995	1999	·1	·5	·9
Panel 01															
Panel Summary		Av + 0·83	Av + 0·62	Av + 0·84			*-----M-*								
Panel 02															
502	F	+1	0												
304	E	+1	0	0·7			1		5	9					
106*	E	0	−1	0·6				1		5					00
109	F	0	+1	0·9				1	5 9						
523	F	+1	+1	0·9			1	5	9						
427	G	+1	+1												
212	E	0	+1	0·9			1	5	9						
335	F	+1	0	0·5							1			05	15
403	G	0	+1	0·6				1	5						10
Panel Summary		Av + 0·56	Av + 0·44	Av 0·74				*----M---*							
Panel 04															
Panel Summary		Av + 0·71	Av + 0·71	Av 0·92			*---- M - -*								
Panel 09															
Panel Summary		Av + 0·78	Av −0.23	Av 0·60			*- ----- M---*								
TRW Summary		Av +0·72	Av +0·38	Av 0·77			*-- --*								

Source: North, H. Q., and Pyke, D. L., (1968)

INTERACTING EFFECTS OF FORECASTS

Delphi as it is currently envisaged takes inadequate account of the interaction of events, and the nature of causative connections. There will be both enhancing and inhibiting effects. Although general assumptions can be imposed upon the entire panel, such as 'No all-out nuclear war to be included', the interactions which are incorporated in given forecasts seldom become manifest. In a pioneering paper Gordon and Hayward have grappled with a methodology for the first time to mitigate the impact of this weakness. A second approach which is developing is gaming.

Suppose that a series of scenarios are forecast, with varying probabilities to occur in 1990. If these are designated, for scenarios S1, S2, S3, . . . Sn, as pS1, pS2, pS3, . . . pSn, the question can be posed: If pSi = 100 per cent, i.e., if scenario i occurs, how if at all do pS1, pS2, pS3, . . . pSn alter? If there is a cross-correlation, the probabilities of these other scenarios will vary either positively or negatively. This type of ramification has been discussed by Parker in the market for conventional fats and oils. The rapid advance of the world synthetic rubber industry during and since the Second World War led to the discovery that a derivative of rosin had very special qualities useful in the emulsifying stage of SBR manufacture. This created an increased demand for rosin, which then became economically recoverable from tall-oil, a by-product of the Kraft paper industry previously treated as waste. The recovery of rosin from tall-oil automatically provides large quantities of fatty acids which needed a market to make the process really viable, and this massive increase in supply affected dramatically the market for conventional fats and oils.

This is an historical example. The problem of future cross-correlations will be readily seen as hazardous territory. None the less, computer applications will undoubtedly soon make it perfectly feasible. Without computer power, a single paradigm as shown in Figure 11 can be developed to accompany the improved conventional approach we have described.

Gaming procedures can be employed to help in the exploration or iteration of the various cross-correlations, and probabilities not just of the direction of movement, but of its magnitude can

Figure 11. *Relationship of a Given Scenario (i) to Eight Other Scenarios*

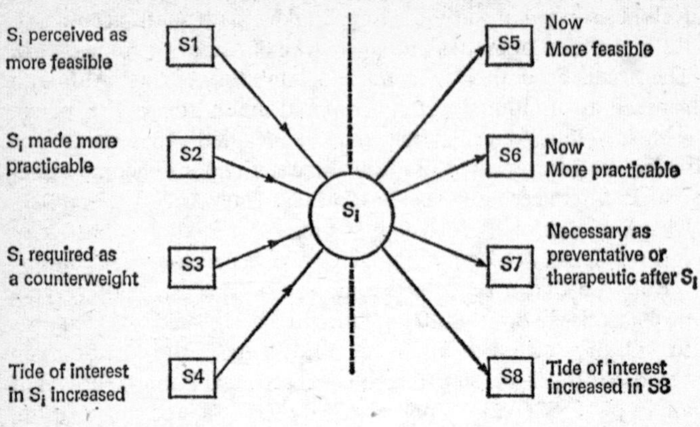

be ascribed. Such games currently involve the verbal exploration of such possibilities in groups, but the intervention of the computer is already happening. Multi-person access to a computer facilitates not just the rapid revision of Delphi method forecasts so that sequential rounds can be conducted within a space of hours rather than over months by post. It also facilitates, through simulation procedures, the exploration of the cross-relationships which have been a problem hitherto.

SOON CHARTING

As with each major TF technique area, we must examine the shortcomings as well. There can be not the slightest doubt that R & D departments and marketing managements, to mention only two of the most sympathetic groups, find Delphi forecasts hard to digest. What precisely does management do with them?

Quite clearly this is something which will be explored in greater detail in the next chapter on Technological Mission Analysis. However, it is appropriate here to consider SOON charting, which was developed by TRW to face up to this very problem after they had completed their Delphi probes. SOON is an acronym for Sequence Of Opportunities and Negatives. Figure 12 reproduces the SOON chart developed at TRW for three-dimensional colour Holographic Films. Such charts display the intervening stages – pre-requisites and any alternatives – before the business can attain any important identified scenario. It is another form of relevance tree analysis, which is included here since it has been specifically deployed following on from the problem of relating scenarios to present-day R & D programmes and budgets. For each and every scenario which a company espouses, this pattern of SOON charting can usefully be developed. Then, but not till then, operating divisions and planners in the short term can begin to identify the technology routes that they must travel to actualize their elegantly formulated scenarios within the time-scales which Delphi has deemed appropriate as well as competitively necessary. The similarity of SOON charts to conventional network diagrams will have not escaped the eye of the seasoned manager. They are a further genus of the ubiquitous decision-tree species.

Figure 12. *A Programme for the Development of 3-D Colour Holographic Films with Some Probable Corollary Developments Indicated (Logic Network)*

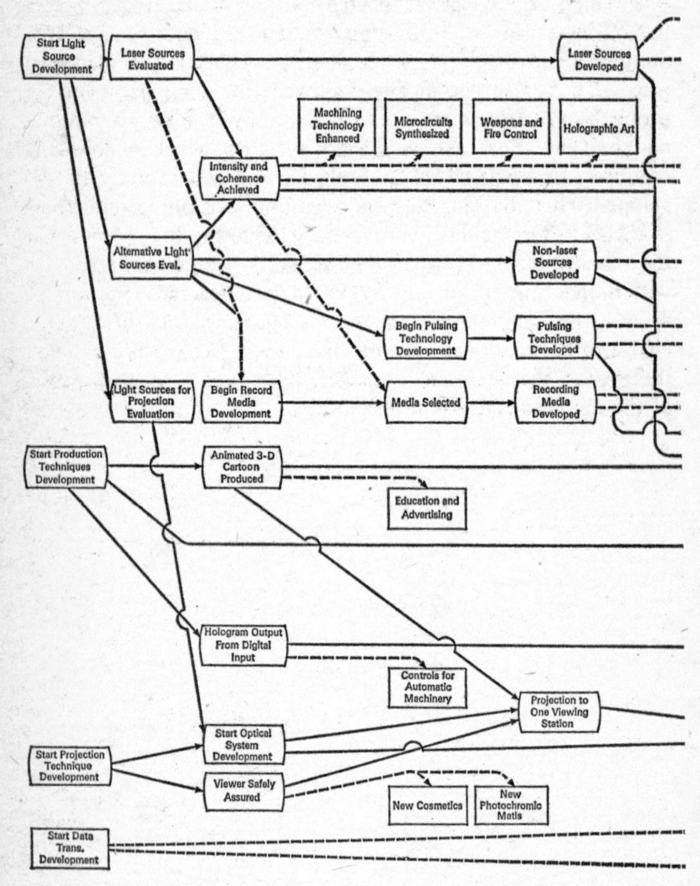

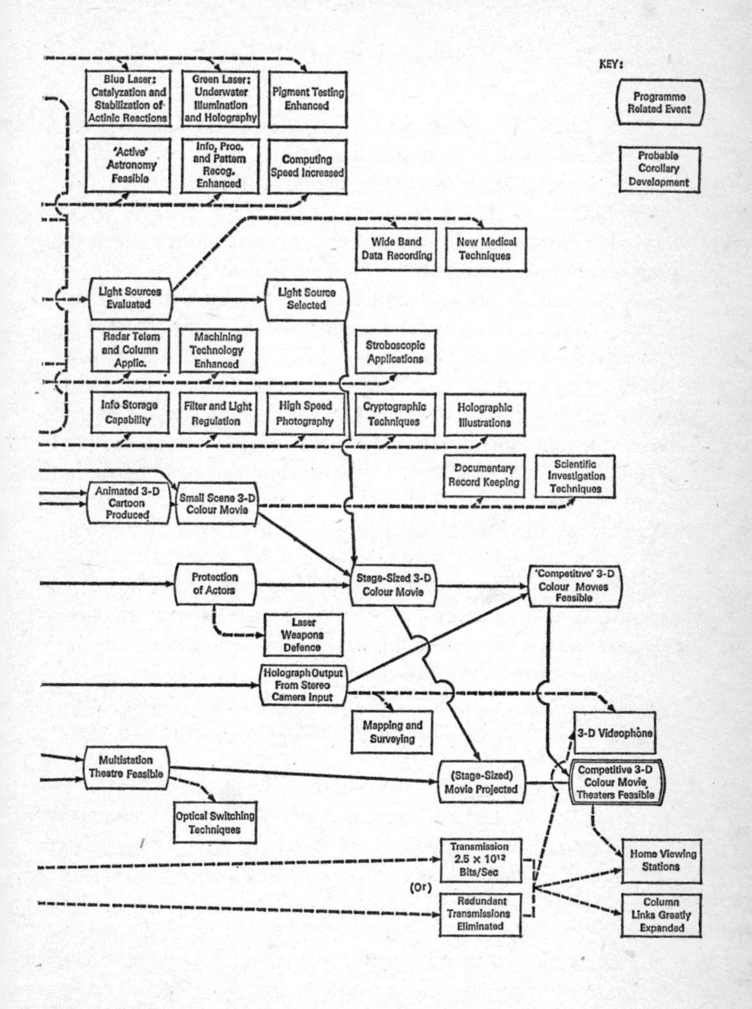
KEY:
Programme Related Event
Probable Corollary Development
Blue Laser: Catalyzation and Stabilization of Actinic Reactions
Green Laser: Underwater Illumination and Holography
Pigment Testing Enhanced
'Active' Astronomy Feasible
Info, Proc. and Pattern Recog. Enhanced
Computing Speed Increased
Wide Band Data Recording
New Medical Techniques
Light Sources Evaluated
Light Source Selected
Radar Telem and Column Applic.
Machining Technology Enhanced
Stroboscopic Applications
Info Storage Capability
Filter and Light Regulation
High Speed Photography
Cryptographic Techniques
Holographic Illustrations
Documentary Record Keeping
Scientific Investigation Techniques
Animated 3-D Cartoon Produced
Small Scene 3-D Colour Movie
Protection of Actors
Stage-Sized 3-D Colour Movie
'Competitive' 3-D Colour Movies Feasible
Laser Weapons Defence
Holograph Output From Stereo Camera Input
Mapping and Surveying
3-D Videophone
Multistation Theatre Feasible
(Stage-Sized) Movie Projected
Competitive 3-D Colour Movie Theaters Feasible
Optical Switching Techniques
Transmission 2.5 × 10¹² Bits/Sec
(Or)
Redundant Transmissions Eliminated
Home Viewing Stations
Column Links Greatly Expanded

CHAPTER 2.5

Technological Mission Analysis

TECHNOLOGICAL MISSION ANALYSIS (TMA) extends our exploration of possible technological futures from a preliminary inventory of actionable technologies right through to potential operationalization of the resultant product or process in its market. The inputs to TMA are all the forecast futures which the full range of TF techniques has identified and which at a preliminary evaluation are in line with corporate objectives. If the company is to carry them further, to examine the full implications of commercialization within its business, a great deal of further analytical rigour is necessary. This means that the firm must look to see whether or not its current or foreseeable resources can achieve the capability levels predicated by the forecasts. Let it be remembered that our forecasts have been framed not from an internal company point of view but in the context of the total competitive technological framework of international business.

TMA is hence an evaluative process in which company capabilities are matched alongside the requirements of the competitive technological environment to see where satisfactory competitive performance is likely to be achieved. It is a coordinated analysis involving integrated contributions from all major functional areas of the business. It demands the generation of alternative technological solutions to forecast futures, and the matching of demand and cost estimates.

Its critical importance to a business is well demonstrated by the attrition rate which it normally affords amongst new ideas brought forward for evaluation. Statistics vary, but most generally suggest that it takes thirty or forty actionable ideas to produce one thoroughly worthwhile commercial venture. This has created specific problems for management. In a large organization, closely concerned with the operational management of today's profit earning products, a continuing series of development fatalities can be dispiriting. More importantly, however, it tends to lead to insufficient attention being paid to securing the future with commercially viable products or processes. Equally, the more routine pattern of management which is effective for current operations frequently jarrs with the radical departures which may quite often be necessary as a company adapts to its future.

The concept therefore arose and is now extensively employed in business, of hiving-off small venture groups or task forces to take complete charge of the evaluation and development of new ideas within the firm. Such groups are inter-functional and devote time solely to the venture or task in hand. It is this interfunctional group concept which TF has extended and which undertakes TMA.

Each group has a mission, an actionable technology. Its task is to evaluate and cultivate that technology as a potentially profitable market offering for the business. Its approach is dynamic and entrepreneurial.

Mission groups are a corporate antidote to anti-entrepreneurialism within the firm, to sclerotic organization structures. They afford the young and the creative the opportunity to facilitate rapid technology-transfer from R & D to the market place at a profit. They are part of the answer big business is offering to the criticism that too much innovation takes place in small firms because the larger firms conspire to prevent change. (This charge, incidentally, misses the point. Large firms seldom conspire, but unless they take careful measures they become creatively impotent.)

Mission groups are usually peopled by rising corporate talent, extremely well-versed in the techniques of modern management technology and management science – graduates in management and capable young technologists. TMA has, therefore, not surprisingly developed into a sophisticated process for the probabilistic quantification of various courses of development within the relevant field of actionable technology. TMA ensures balanced consideration of all relevant corporate aspects. It eliminates the bootleg project in the laboratories about which management knows little; the bottomless purse project that took four times as long, and cost six times as much as expected; the product, 25,000

Figure 1. *The Sequence of Technological Mission Analysis*

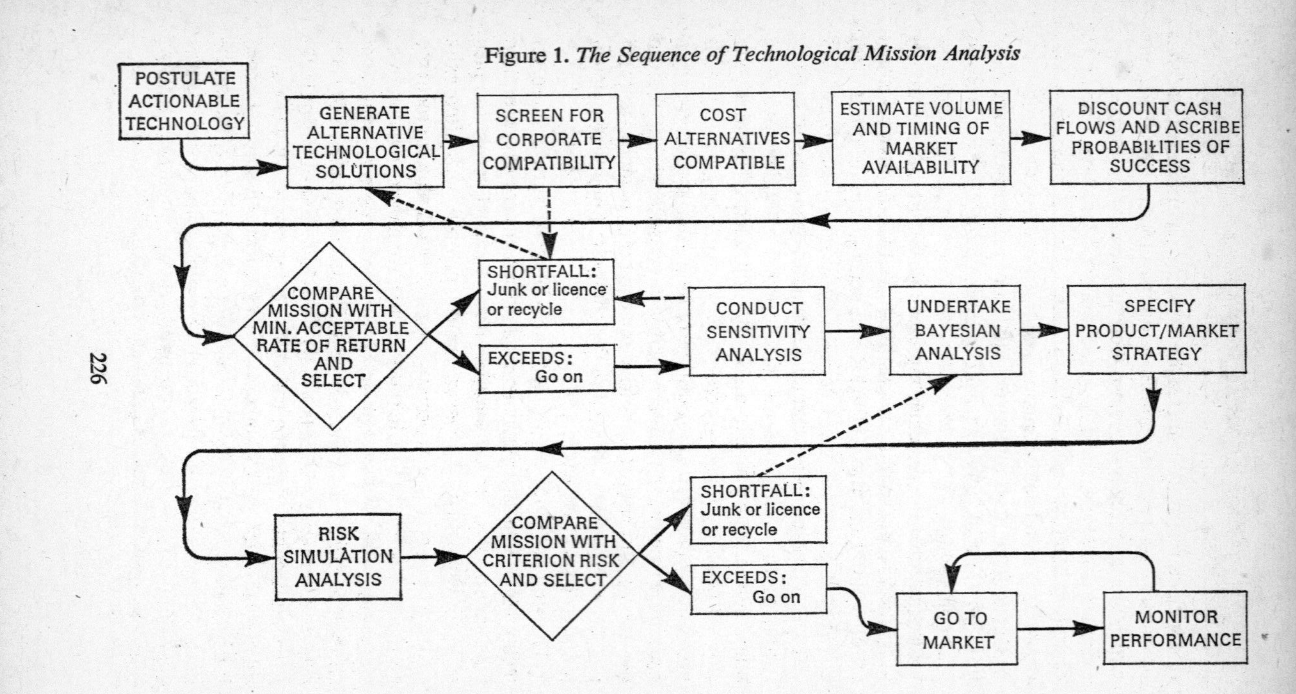

of which came back from customers with bugs in them; and the product which had all profits and sales engineered out of it but which immediately became a collector's item.

Figure 1 displays the TMA sequence. We shall be looking at each stage in turn and the particular techniques which are deployed It will be useful here to explore verbally a postulated future which has been generated by, in this instance, scenario writing, as it undergoes TMA. Such an actionable technology might well be dirt repelling fabric.

The first step is undertaken within the R & D department of the business. There, creative scientists will generate as many alternative technological ways of achieving such a fabric as possible. One possibility would be to treat existing fabrics with a chemical in order to cause it to reject dirt on contact. A second line of investigation could well be in the development of a new synthetic fabric whose basic make up was resistant to dirt. Given these two alternatives, they must be screened for their compatibility with present and potential corporate resources. An absence of major manufacturing capacity might well lead to a preference for the development of spray technology; on the other hand, if the company was a major manufacturer of synthetic fibres, then a new fabric might be compatible.

The compatibility screen will not be used to eliminate all technological alternatives, but only those which are not compatible with present and future resource anticipations. Even in the face of such a constraint, however, the opportunity will often exist to sub-contract some aspects of technological development or even to buy-in knowledge in the field to shorten development time. Company mergers and acquisitions may also be employed as a strategy to extend the concept of resource compatibility. For each arrayed alternative, costings will then be made. Especially important in this context are the estimates of the minimum period of time over which development can be successfully accomplished for each alternative. This timing aspect will have crucial importance for marketing's assessment of total demand or market availability as also will cost. Demand estimates will be at a range of prices in given places at a given time interval.

Financial members of the TMA group can now add to their cost outflows in development the hypothesized cash flows into the business as sales accrue. This will be accomplished by the modification of original estimates of *total* market demand by probabilistic assessments of the proportion which our company might expect to obtain. When these net cash flows are discounted over their time cycles, we are in a position to make a comparison with the minimum acceptable rate of return to the business. If we fall short of this, we must either discard our idea, licence it to others, or recycle it back to the generation of alternative technological solutions step. If the minimum acceptable rate of return on investments is exceeded, then TMA takes an added precautionary step in what is termed sensitivity analysis of the outcome so far. The assumptions made about each contributing element – the time scale for development, the cost estimates and market forecasts – are all tested to determine how robust the anticipated rate of return on the mission will be if those basic assumptions turn out to be incorrect. The extent of such tolerance in the conclusion before it falls short of the minimum acceptable rate of return will be determined. If the conclusion is insufficiently robust, then that line of development may well be rejected for the mission. At this stage, final selection of the technological alternative to pursue in fulfilment of the mission is sometimes made if more than one alternative remains acceptable. Normally the alternative approach affording the maximum potential rate or return on investment will be chosen.

However, a second analytical phase is often incorporated in TMA, involving Bayesian and Risk Simulation Analyses. These employ probabilistic approaches to the outcomes forecast in order to attempt to evaluate additional information for reducing uncertainty and they afford the opportunity for corporate risk-preference criteria to be encompassed. They will normally include the employment of marketing research techniques for product concept and placement tests and often test pilot experimental marketing. In the case of our fabric, it might well be made up into garments for use by selected panels of customers and their attitudes and reactions would be evaluated. Provided all goes well, the company would now be ready to go to market to monitor product performance and adjust its offering as necessary.

Against this overall descriptive background, let us now turn to look in detail at the analytical quantification procedures employed.

SCREENING FOR CORPORATE COMPATIBILITY

The creative generation of alternative technological routes to the fulfilment of a technological mission will not be impeded by screening for corporate compatibility. It none the less constitutes the first major constraint on the possible pursuit of otherwise preferable routes. Each business will, at any point in time, have a given pattern of resource allocation with synergetic potential available therefrom. Synergy is the concept of $1+1 = 3$, the realization in business terms that existing patterns of resource allocation can afford considerable benefits from carry-over effects from other areas of activity. An extant capacity for spray technology or a substantial level of technological know-how in spray treatments would both afford a synergetic benefit to our dirt-repellant spray alternative if we chose to adopt it over and above the development of a new fabric. None the less, synergy search must be undertaken with due caution. Synergy may lead the myopic to follow courses of action which expose them to severe competitive disadvantage from others following a more hazardous but, if successful, greatly more rewarding avenue of development.

Each business can identify those factors which it deems especially important and against which it wishes to assess technology-route compatibility. The classic listing which encompasses almost every facet of an organization's resources was prepared by O'Meara, and is given in Figure 2. This listing of factors and sub-factors is the starting point for a comparative quantification of the alternatives. Provided a predetermined score is attained, the technology-route may be deemed compatible. Four main factors are normally identified – marketability, durability, productive ability and growth potential.

In Figure 3, a typical analysis is presented. For each level of compatibility, a weight has been suggested:

Figure 2. *Corporate Compatability Screening Factors (After O'Meara)*

	Very Good	*Good*	*Average*	*Poor*	*Very Poor*
I. MARKETABILITY					
A. *Relation to present distribution channels*	Can reach major markets by distributing through present channels.	Can reach major markets by distributing mostly through present channels, partly through new channels.	Will have to distribute equally between new and present channels, in order to reach major markets.	Will have to distribute mostly through new channels in order to reach major markets.	Will have to distribute entirely through new channels in order to reach major markets.
B. *Relation to present product lines*	Complements a present line which needs more products to fill it.	Complements a present line that does not need, but can handle, another product.	Can be fitted into a present line.	Can be fitted into a present line but does not fit entirely.	Does not fit in with any present product lines.
C. *Quality/price relationship*	Priced below all competing products of similar quality.	Priced below most competing products of similar quality.	Approximately the same price as competing products of similar quality.	Priced above many competing products of similar quality.	Priced above all competing products of similar quality.
D. *Number of sizes and grades*	Few staple sizes and grades.	Several sizes and grades, but customers will be satisfied with few staples.	Several sizes and grades, but can satisfy customer wants with small inventory of nonstaples.	Several sizes and grades, each of which will have to be stocked in equal amounts.	Many sizes and grades which will necessitate heavy inventories.

E. *Merchandisability*	Has product characteristics over and above those of competing products that lend themselves to the kind of promotion, advertising, and display that the given company does best.	Has promotable characteristics that will compare favourably with the characteristics of competing products.	Has promotable characteristics that are equal to those of other products.	Has a few characteristics that are promotable, but generally does not measure up to characteristics of competing products.	Has no characteristics at all that are equal to competitors' or that lend themselves to imaginative promotion.
F. *Effects on sales of present products*	Should aid in sales of present products.	May help sales of present products; definitely will not be harmful to present sales.	Should have no effect on present sales.	May hinder present sales some; definitely will not aid present sales.	Will reduce sales of presently profitable products.
II. DURABILITY A. *Stability*	Basic product which can always expect to have uses.	Product which will have uses long enough to earn back initial investment, plus at least 10 years of additional profits.	Product which will have uses long enough to earn back initial investment, plus several (from 5 to 10) years of additional profits.	Product which will have uses long enough to earn back initial investment, plus 1 to 5 years of additional profits.	Product which will probably be obsolete in near future.
B. *Breadth of market*	A national market, a wide variety of consumers, and a potential foreign market.	A national market and a wide variety of consumers.	Either a national market or a wide variety of consumers.	A regional market and a restricted variety of consumers.	A specialized market in a small marketing area.
C. *Resistance to cyclical fluctuations*	Will sell readily in inflation or depression.	Effects of cyclical changes will be *moderate*, and will be felt *after* changes in economic outlook.	Sales will rise and fall with the economy.	Effects of cyclical changes will be *heavy*, and will be felt *before* changes in economic outlook.	Cyclical changes will cause extreme fluctuations in demand.

	Very Good	*Good*	*Average*	*Poor*	*Very Poor*
D. *Resistance to seasonal fluctuations*	Steady sales throughout the year.	Steady sales – except under unusual circumstances.	Seasonal fluctuations, but inventory and personnel problems can be absorbed.	Heavy seasonal fluctuations that will cause considerable inventory and personnel problems.	Severe seasonal fluctuations that will necessitate layoffs and heavy inventories.
E. *Exclusiveness of design*	Can be protected by a patent with no loopholes.	Can be patented, but the patent might be circumvented.	Cannot be patented, but has certain salient characteristics that cannot be copied very well.	Cannot be patented, and can be copied by larger, more knowledgeable companies.	Cannot be patented, and can be copied by anyone.
III. PRODUCTIVE ABILITY					
A. *Equipment necessary*	Can be produced with equipment that is presently idle.	Can be produced with present equipment, but production will have to be scheduled with other products.	Can be produced largely with present equipment, but the company will have to purchase some additional equipment.	Company will have to buy a good deal of new equipment, but some present equipment can be used.	Company will have to buy all new equipment.
B. *Production knowledge and personnel necessary*	Present knowledge and personnel will be able to produce new product.	With very few minor exceptions, present knowledge and personnel will be able to produce new product.	With some exceptions, present knowledge and personnel will be able to produce new product.	A ratio of approximately 50–50 will prevail between the needs for new knowledge and personnel and for present knowledge and personnel.	Mostly new knowledge and personnel are needed to produce the new product.

C. *Raw materials' availability*	Company can purchase raw materials from its best supplier(s) exclusively.	Company can purchase major portion of raw materials from its best supplier(s), and remainder from any one of a number of companies.	Company can purchase approximately half of raw materials from its best supplier(s), and other half from any one of a number of companies.	Company must purchase most of raw materials from any one of a number of companies other than its best supplier(s).	Company must purchase most or all of raw materials from a certain few companies other than its best supplier(s).
IV. GROWTH POTENTIAL					
A. *Place in market*	New type of product that will fill a need presently not being filled.	Product that will substantially improve on products presently on the market.	Product that will have certain new characteristics that will appeal to a substantial segment of the market.	Product that will have minor improvements over products presently on the market.	Product similar to those presently on the market and which adds nothing new.
B. *Expected competitive situation – value added*	Very high value added so as to substantially restrict number of competitors.	High enough value added so that, unless product is extremely well suited to other firms, they will not want to invest in additional facilities.	High enough value added so that, unless other companies are as strong in market as this firm, it will not be profitable for them to compete.	Lower value added so as to allow large, medium, and some smaller companies to compete.	Very low value added so that all companies can profitably enter market.
C. *Expected availability of end users*	Number of end users will increase substantially.	Number of end users will increase moderately.	Number of end users will increase slightly, if at all.	Number of end users will decrease moderately.	Number of end users will decrease substantially.

Figure 3. *Technology Route Compatibility Screening*

Compatibility Sub-factors	*Sub-factor Weight*	*Compatibility Level Weight*					*Expected Level Weight*	*Contribution to Total Expected Compatibility*
		V. Good	Good	Average	Poor	V. Poor		
Column (i)	Column (ii)	(10)	(8)	(6)	(4)	(2)		
MARKETABILITY								
Relation to present channels	5	0·9	0·1	0·0	0·0	0·0	9·8	49·0
. . . to present product lines	8	0·2	0·2	0·4	0·2	0·0	6·8	54·4
Q/Price	10	0·2	0·4	0·2	0·1	0·1	7·0	70·0
Sizes and grades	3	0·6	0·2	0·2	0·0	0·0	8·8	26·4
Merchandisability	1	0·2	0·8	0·0	0·0	0·0	8·4	8·4
Effect on sales	10	0·5	0·4	0·1	0·0	0·0	8·8	88·0
DURABILITY								
Stability	5	0·0	0·0	0·3	0·4	0·3	4·0	20·0
Breadth	2	0·1	0·2	0·4	0·3	0·0	6·2	12·4
Resistance to cycles	2	0·0	0·0	0·4	0·6	0·0	4·8	9·6
. . . seasonal fluctuations	8	0·8	0·2	0·0	0·0	0·0	9·6	76·8
Exclusivity of design	8	0·4	0·6	0·0	0·0	0·0	8·8	70·4
PRODUCTIVE ABILITY								
Equipment necessary	6	0·5	0·4	0·1	0·0	0·0	8·8	52·8
Know how, etc.	5	0·8	0·2	0·0	0·0	0·0	9·6	48·0
Materials availability	10	0·4	0·6	0·0	0·0	0·0	8·8	88·0

GROWTH POTENTIAL								
Place in market	4	0·2	0·3	0·5	0·0	0·0	7·4	29·6
Value added	3	0·4	0·5	0·1	0·0	0·0	8·6	25·8
Market availibility	10	0·5	0·3	0·2	0·0	0·0	8·6	86·0
				Total Expected Compatibility of Technology Route				815·6

Highly compatible	10
Good compatibility	8
Average compatibility	6
Poor compatibility	4
Very poor compatibility	2

These absolute weights are arbitrarily determined but are hopefully representative of the cardinal ranking of compatibility factors. For each sub-factor in this screening process, the relevant executives will ascribe probabilities to each level of factor compatibility. These probabilities can be ascribed solely by the venture group undertaking the TMA but are more likely to be arrived at in terms of a Delphi-type consensus of well-informed experts in the area. The total weight given to each sub-factor, representing its relative order of importance to the business, is given in column (ii). This weight is multiplied by the sum of the compatibility-level probability and its associated weight (i.e. expected level weight) to provide the final column, contribution to total expected compatibility. The sum of this final column provides, here, an overall compatibility for the technological route in question of 815.6. This index may be compared both with the compatibility norm and with the alternative routes under consideration.

The concept of a compatibility norm is complex. It is usually an amalgam of total scores of previously successful technology routes of which the company has experience.

TIMING AND COSTING OF ALTERNATIVE TECHNOLOGY ROUTES

Determination of compatibility with corporate resources and dispositions must precede any formal time analysis for R & D programmes and the costing of alternative budget rates. The procedure now adopted involves analysis to develop the data indicated in Figure 4.

Three probability rates for completion are indicated. Only the maximum and intermediate effort are deemed worthwhile, since minimal effort seems almost certain to fail to provide any viable solution. It is crucial to remember at this juncture that soonest, i.e., maximum, R & D effort is not necessarily the optimum timing for the business. Not only may it involve higher absolute expenditure but the timing of market availability must be incorporated into the analysis. An analogous approach to market availability in terms of market development budgets can be described, as in Figure 5.

Without any effort, market demand will unfold before the business – the lower curve. A moderate probability that market demand already exists is indicated. Absolute certainty about market demand, however, at a requisite level, will not be available on current estimation, for seven or eight years even if we make a maximum level of marketing investment to stimulate sales.

In each instance, TMA demonstrates the discretion which exists in timing technological and market availability, and the relative nature of cash flows which would accompany the course of action. The optimal moment for introduction can be discerned as that moment at which the total time related costs of introduction are minimized. This is determined by an extension of a technique known as discounted cash flow (dcf) analysis. First dcf will be briefly described and then the extension for timing can be made.

Discounting of cash flows is the mirror image of compound interest rates, a familiar enough concept. Let r equal the annual rate of interest payable, then at the end of n years, a sum of £A invested at r per cent per annum will have grown to B.

$$B = A\,(1+r)^n$$

To determine the sum which needs to be invested now (A), which will yield B in n years, at r per cent per annum, we simply manipulate the compound interest equation to yield:

$$A = \frac{B}{(1+r)^n}$$

Figure 4. *Successful Completion of R & D with Differing Budget Levels*

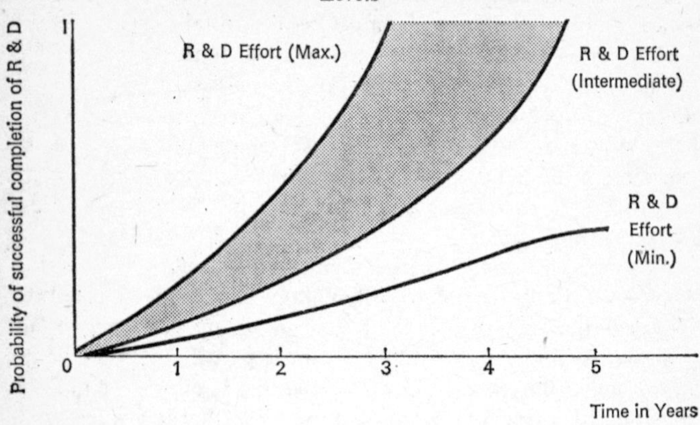

Figure 5. *Successful Development of Requisite Level of Market Demand with Differing Budget Levels*

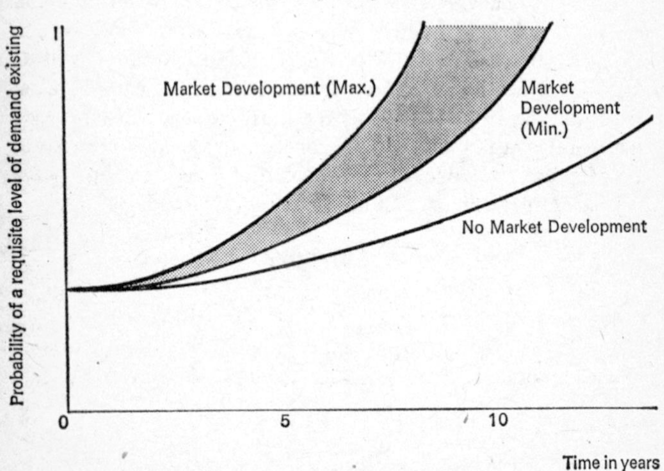

Figure 6. *Comparative Cash Flow Analysis for Two Alternative Technological Routes (£000)*

	Years After Start									
	0	1	2	3	4	5	6	7	8	9
Technology Route A										
Revenue	0	2,000	5,000	20,000	5,000	0	0	0	0	0
Costs	10,000	2,000	2,000	2,000	2,000	0	0	0	0	0
Net Cash Flow	−10,000	0	3,000	18,000	3,000	0	0	0	0	0
Technology Route B										
Revenue	0	2,000	10,000	15,000	15,000	15,000	2,000	2,000	2,000	0
Costs	20,000	3,000	3,000	4,000	4,000	4,000	1,000	1,000	1,000	0
Net Cash Flow	−20,000	−1,000	7,000	11,000	1,000	11,000	1,000	1,000	1,000	0

This expression for A is what in dcf jargon is termed the *present value of B*. In other words, we have taken cognizance of the fact that £1 today is worth more than £1 in n years' time. Inflation apart (and we need to allow for that), the sum of £1 now can be invested to earn interest during those n years. Hence, less than £1 today will be worth £1 n years hence. Why do this? Because for different alternatives, the value of n will differ – for some products it may be five years, for others fifteen years before cash flows reach particular levels. By reducing all alternatives to their present expected value we have a fair yardstick for their comparison. Figure 6 shows two different patterns of cost and revenue flows for two distinct technology routes. In both cases, heavy R & D and market development expenditures are envisaged at the outset; in the case of technology route A a short product life cycle is expected. Route B is expected to offer a longer life cycle. A similar analysis could be undertaken to compare maximum with other levels of effort in R & D and market development.

Applying the equation already developed, with a 10 per cent discount rate, technology route A has a present value of £8 million as compared with route B's £9 million. *Prima facie*, route B seems to be preferable. Both certainly had positive present values, and were hence within the range of viable investment goals for the business. A considerably different pattern of cost and revenue flows could be derived for different levels of effort. The simple example in Figure 6 has assumed just one level of effort. A search can be mounted to arrive at the optimum point of introduction of products via technology routes; that is to say the point in time where present value is maximized.

This optimum present value will occur when the incremental gain by delay is equivalent to the incremental loss by delay, i.e. a firm will delay so long as

$$\frac{A}{(1+r)^t}\left(1-\frac{1}{(1+r)}\right)>\frac{Q(t+1)}{(1+r)^{t+1}}$$

where

$\frac{Q(t+1)}{(1+r)^{t+1}}$ = loss in present value resulting from delay of introduction from t to t+1, i.e., net cash flow divided by the discount factor.

A = initial investment required.

r = opportunity rate of return expected.

The global nature of the timing problem for introduction of a new technology-based market offering is demonstrated in Figure 7. There two functions are indicated. The cost of going too slowly embraces all positive revenue flows into the business which would have accrued if the market had been entered and most particularly the higher cost of later penetration if competitors become entrenched. The cost of going too rapidly, however, will include not just the cost of capital tied up over time in the project but also the higher likelihood of making errors if one is first to market. These are to some extent *counteracted* by the benefits of an entrenched position in the market early on, particularly permitting choice of distributors to be made.

Figure 7. *The Timing of Market Introduction*

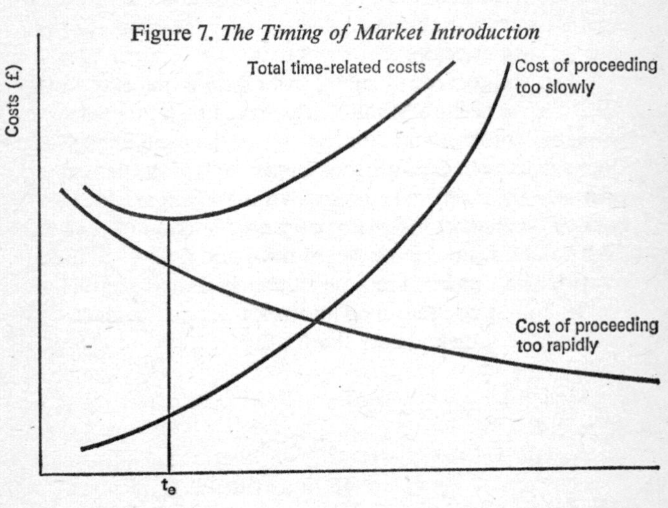

The optimum time for market entry is shown at t_e, where total time-related costs are at their minimum.

SIMULATION OF CASH FLOW AND RATE OF RETURN

We have so far taken a somewhat casual look at the potential profitability of the market offering which might result from the technological mission. Fairly rigid assumptions have been made about costs. It is necessary to look at cost and revenue flows on precisely the same probabilistic basis as we first used to screen for compatibility (Figure 3). One popular method has been dubbed PIE (Product Innovation Evaluator). Its sequence is described in Figure 8.

The venture group's first task is to secure estimates of all the significant parameters - i.e., annual unit sales, selling price, unit cost, R & D start-up expenses and capital investment cash flows. These estimates are not presented in the form of a rigid single point datum, but rather as a probability distribution. X represents the value of each variable and σ its standard deviation. That is a measure of its dispersion.

PIE then commences to simulate the future outcome for the market offering. Using a Monte Carlo procedure (in the same way as ERNIE selects Premium Bond winners), the simulation selects random values from each distribution for each of the parameters, according to the estimated chances that the selected value will be realized in the market place. For each combination of randomly selected values, the discounted cash flows and rates of return are computed. This process is repeated and the mean and variance of annual cash flows and return on investment are determined. These twin outcomes are shown at the bottom right-hand corner of Figure 8.

The expected value or mean cash flow is plotted with 1σ shown to either side of it. This measures the zone within which we are reasonably certain the answer will fall. We can extend that zone of reasonableness to higher levels of confidence by plotting 2σ or 3σ.

Figure 8. *Product Innovation Evaluator* (*Pie*)

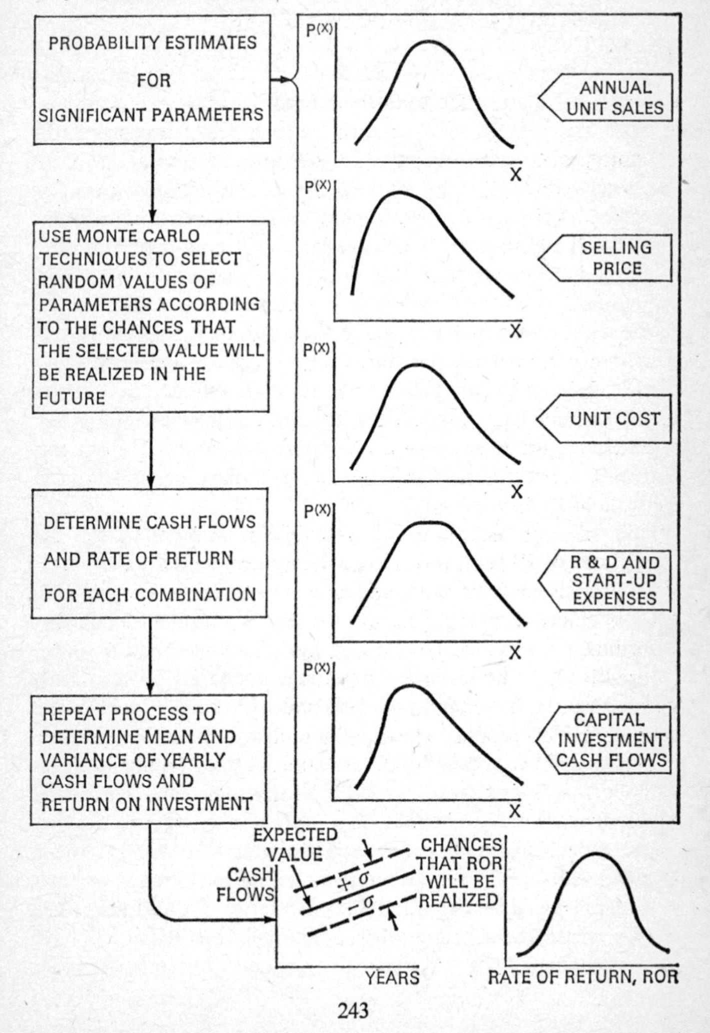

P(X) = CHANCES THAT VALUE WILL BE REALIZED
X = VALUES OF VARIABLE
σ = STANDARD DEVIATION
PROBABILITY ESTIMATES FOR SIGNIFICANT PARAMETERS
USE MONTE CARLO TECHNIQUES TO SELECT RANDOM VALUES OF PARAMETERS ACCORDING TO THE CHANCES THAT THE SELECTED VALUE WILL BE REALIZED IN THE FUTURE
DETERMINE CASH FLOWS AND RATE OF RETURN FOR EACH COMBINATION
REPEAT PROCESS TO DETERMINE MEAN AND VARIANCE OF YEARLY CASH FLOWS AND RETURN ON INVESTMENT
P(X)
X
ANNUAL UNIT SALES
SELLING PRICE
UNIT COST
R & D AND START-UP EXPENSES
CAPITAL INVESTMENT CASH FLOWS
EXPECTED VALUE
CASH FLOWS
+σ
−σ
YEARS
CHANCES THAT ROR WILL BE REALIZED
RATE OF RETURN, ROR

Figures 9(a) and (b) display in further detail the implications of the PIE output. It will be seen that estimates are 100 per cent certain that the rate of return will not exceed 30 per cent nor fall below 23 per cent.

The advantages of PIE will now be apparent. It enables the business to evaluate the expected rate-of-return for the market offering and the probabilities that it will be achieved (that is to say the risk) as well as indicating the cash flow requirements in the early years and the range of uncertainties associated with all the significant cost and revenue parameters. In particular, it puts R & D cost into a true perspective. It will often be substantially lower than the innovation expenses within the business and within the market place.

Sensitivity analysis, it will be recalled, is used to determine the parameters and distributions to which output is most sensitive. Its purpose is to permit the business to perceive just exactly where the greatest uncertainty lies and thereby to facilitate the collection of additional data if appropriate to reduce such uncertainty. If a parameter is found to be particularly critical for the PIE output, further data collection or a second look at managerial estimates must be taken.

The main method of sensitivity analysis available to test the robustness of PIE output is experimentation, either on a judgemental basis or with a full scale factorial design. Judgemental experimentation will test for sensitivity to a range of logically anticipated change in the distributions of values for each parameter. It keeps the testing at a manageable level. To conduct a comprehensive factor analysis of three distributions of each of the five parameters would involve 3^5 or 243 separate simulations.

Finally, PIE has presented the data necessary for the business to match its risk preferences with the risk estimated as within the technological mission overall. The particular mission to develop a market offering which is under investigation will be plotted within the trade-off chart of senior management for the business which is shown as Figure 10. From the PIE simulation the mean expected rate of return and the standard deviation of its estimates will have been computed. It is hypothetically shown in Figure 10 at point

Figure 9(a). *Typical Pie Plots of Cash Flow for Technological Mission Analysis*

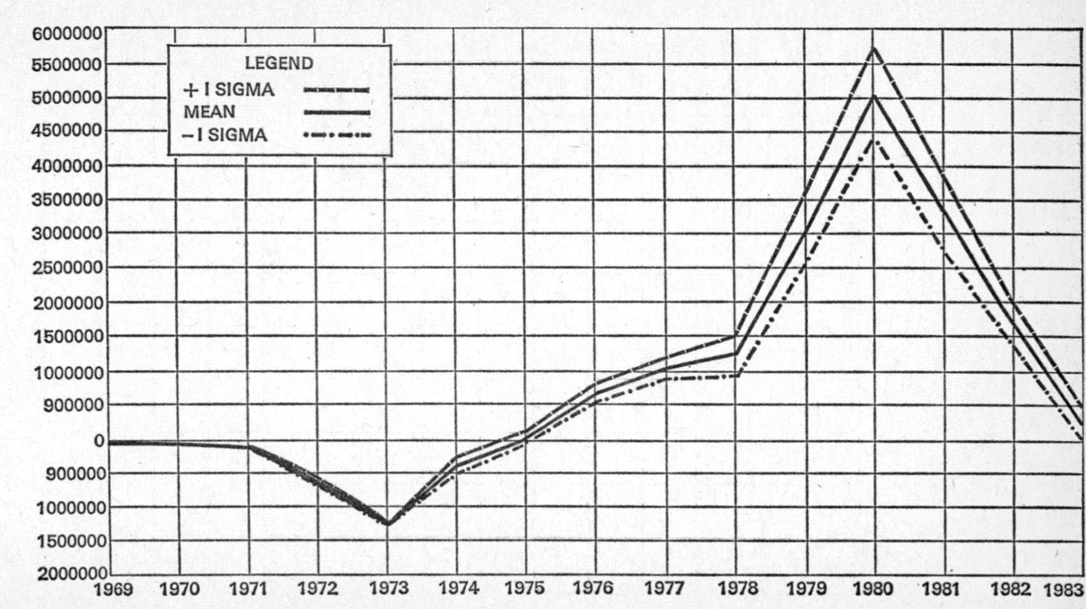

Figute 9(b). *Typical Pie Plots of Rate-of-Return for Technological Mission Analysis*

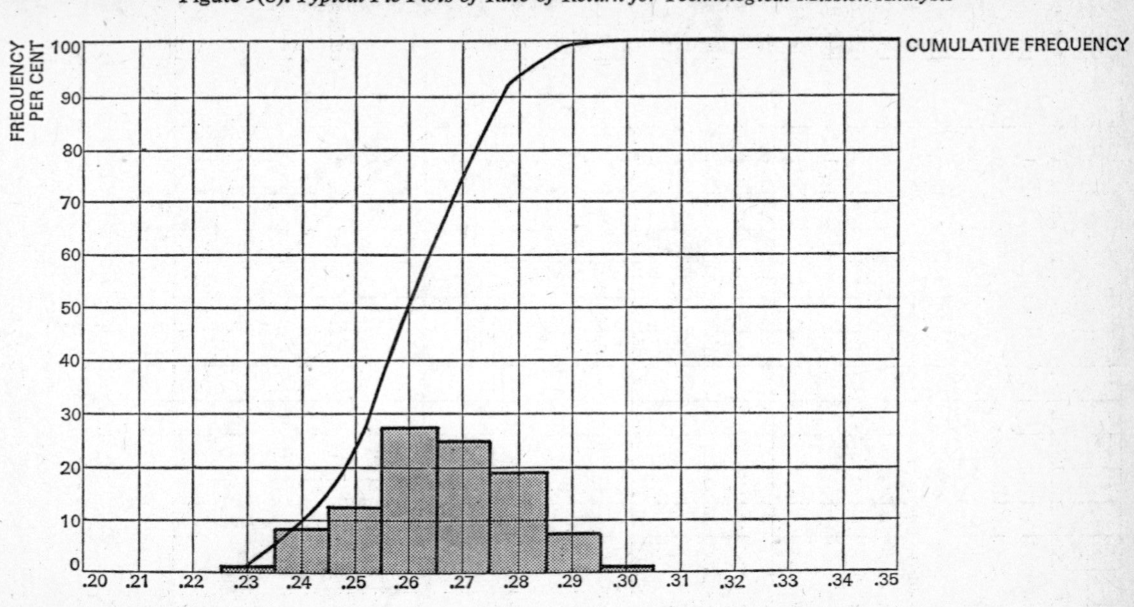

J, indicating that although quite a wide scope for risk exists, the mean expected rate of return is sufficient to counterbalance it. Had the outcome been at G, then management would avert the risk.

Such analysis does, of course, suffice to compare two different technology routes. J and G could have been such routes. So could L – a route with much less risk but for fewer rewards. Which of these two acceptable technology routes, J or L, will be followed will often depend on the recent history of innovatory activity within the firm, and the size and criticality of the investment involved for the business. There is nothing sacred about the trade-off chart shown, and that is why no specific figures have been inserted. It will change from time to time. A gambling or a desperate management may increase the slope of the curve, whereas a

Figure 10. *PIE Trade-off Between Risk and Rate of Return*

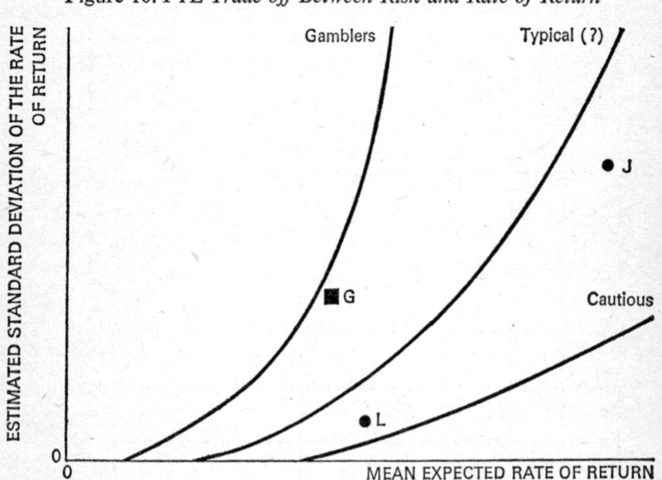

cautious group of executives might reduce it. These two alternatives to typicality are shown by the additional lines on Figure 10. It will be seen that they would either encompass all technology routes that are shown or exclude them all.

EXPECTED PRESENT VALUES AND THE VALUE OF ADDITIONAL INFORMATION

One way of reducing risk is to collect further relevant information about the market place into which the market offering will be launched. The technological mission analysts have at their disposal here the powerful technique called Bayesian Decision Theory. It is perhaps best illustrated in the context of the test marketing activity for the business. The Rev. Thomas Bayes was an eighteenth-century British thinker who, in simplified terms, evolved an approach to decision-making which works with the economic costs of alternative courses of action, the prior knowledge of the decision-maker, and formal modifications of such judgement as and when additional information becomes available. It is a sequential process of analysis, involving three analytical elements – prior analysis, posterior analysis and pre-posterior analysis.

Let us assume in this context that a business is contemplating a single technology route of development and wishes to examine the potential in two distinctly separate international markets – say Japan and Canada.

(i) *Prior Analysis* involves choosing amongst alternative courses of action where no data exists for evaluation. The TMA group is called upon to use its prior judgement (based on its experience and other research) to attribute probabilities to the range of alternatives occurring given each possible course of action. Pay-offs reflect the expected financial value of each outcome deemed possible.

(ii) *Posterior Analysis* enables the TMA group to interpret data as it becomes available in order to revise its prior judgements.

(ii) *Pre-posterior Analysis* enables the TMA group to evaluate the worth of additional information before purchase in terms of its anticipated effect on reducing the costs associated with uncertainty. Further, this approach can be used to evaluate relative efficiencies of differing marketing research plans and sample sizes.

The specific pattern of network analysis known as decision trees can fruitfully be employed in exposing the implications of the alternatives, their probabilities and discounted expected values.

The Two Markets

Let us now examine the two markets in detail – Japan and Canada. It is into these that we are contemplating our launch. Neither market will make millionaires overnight of the business's shareholders, but both seem to have pleasant prospects of success – Canada more so than Japan. For each market, the TMA group estimates probabilities of various market shares – say 5 per cent, 10 per cent and 15 per cent – being the true state which will occur. These estimates are shown in Figure 11, and are the prior probabilities along with the expected value of each course of action.

The final column of Figure 11 is the product of the implicit profit for any particular outcome of broadscale share in the market place and the mean probability estimate of its occurrence. The prior expected values of the two markets are £680,000 and £10,400,000 respectively, or the sum of the last column's profit consequences. We have assumed that the three broadscale shares are exclusive and all-embracing outcomes. In practice, many more outcomes may be envisaged as possible and hence included in such analysis.

Figure 11. *Bayesian Prior analysis for Markets in Japan and Canada*

Forecast Broadscale Market Share per cent	*Implicit Discounted Profit (£m)*	*Estimated Mean Probability*	*Profit Consequence (£)*
JAPANESE MARKET			
15	8	0·2	1·6
10	2	0·4	0·8
5	–4·3	0·4	–1·72
Expected Value of Japanese Market = £0·68m.			
CANADIAN MARKET			
15	30	0·3	9
10	12	0·5	6
5	–23	0·2	–4·6
Expected Value of Canadian Market = £10·4m.			

A measure of risk in going ahead in each market has now also emerged. In Japan we estimate that we have a 0·4 chance of losing £4·3m.; £1·72m. is at stake. In Canada at stake we have £4·6m. These values are termed the expected value of perfect information (EVPI) and are the maximum sums of money we feel are at risk. These are theoretical maxima for information gathering budgets. In view of the size of the EVPI, the TMA group may well determine that a test market exercise is worthwhile, and a budget of £250,000 for each country is tentatively proposed. How certain can we be that if such tests were conducted, and if given market shares materialized on those tests, they would in fact accurately reflect the true state of nature, i.e., the real broadscale market share? We certainly cannot anticipate the collection of 100 per cent accurate information. Test marketing is not that accurate a technique and, in any event, total accuracy is unnecessarily expensive to the business. In the light of previous experience, let us suppose that the TMA group feels a budget for a test market of £250,000 will give reliability as a predictor of broadscale shares as indicated in Figure 12.

Figure 12. *Probabilities of Test Market Shares Materializing in the Light of Given Broadscale Market Shares in Japan or Canada*

	Sustained Test Market Penetration (per cent)		
Broadscale Market Shares (per cent)	Over 15 per cent (TS_1)	10–15 per cent (TS_2)	Below 10 per cent (TS_3)
15 (BS_1)	0·9	0·05	0·05
10 (BS_2)	0·1	0·8	0·1
5 (BS_3)	0·0	0·15	0·85

There is no reason why these probabilities should be identical for both Japan and Canada. Different qualities of research facilities in each country would undoubtedly be reflected in such ascribed probabilities. The diagonal values (0·9, 0·8, 0·85) indicate the credence which the TMA feels can realistically be imparted to test market outcomes. If these seem low, let me reassure the reader they are in fact somewhat high by popular experience. Despite Galbraith's beliefs, business does *not* have a particularly powerful tool for predicting behaviour in this area.

Different values of test-market budget will, of course, also affect the level of credence given, but there is no straight-line increase in accuracy with budget increases.

The next computational stage required yields the probabilities as shown now (at time *t*) of the joint occurrence of given test and broadscale shares. From these the posterior probabilities are calculated. The results are given in Figures 13 and 15 for Japan and Canada.

The probability of a joint occurrence is the product of the prior probability and the conditional probability. Hence, in Figure 13, 0·18 probability of the joint occurrence (TS_1 given BS_1) is the product of 0·2 (BS_1 Figure 11) and 0·9 (TS_1/BS_1 Figure 12). The marginal totals are equal to the prior probabilities in Figure 11.

Posterior probabilities are then computed by dividing each probability of joint occurrence by the column totals, i.e., the likelihood of the test market brand shares materializing. In Figure 13 posterior probability 0·818 (TS_1/BS_1) is the outcome of:

$$\frac{0{\cdot}18\ (TS_1 \text{ given } BS_1)}{0{\cdot}22\ (\Sigma TS_1)}.$$

We have already noted that the expected value of perfect information would be saving the losses which would be incurred if BS_3 materialized in either Japan or Canada. These were £1·72m. and £4·6m. respectively. The final computations are designed to establish whether the discounted value now (at time *t*) of test-market information is increased to a greater extent than the £0·25m. incurred in test markets.

Posterior and pre-posterior analyses, which yield this information, are demonstrated in Figures 14 and 16. Figure 14 for Japan shows an increased value of introduction now, if test-market data is acquired, of £1·983m. – £0·68m. = £1.303m. at a cost of £0·25m. If we abide by the decision rules implicit in our analysis then the test market should be undertaken.

Figure 13. *Japanese Market*

Broadscale Brand Shares	*Probability of Joint Occurrence of Test and Broadscale Brand Shares*			*Marginal Total*	*Posterior Probabilities Taking Cognizance of Test Data*		
	TS_1	TS_2	TS_3		TS_1	TS_2	TS_3
BS_1	0·18	0·01	0·01	0·2	0·818	0·026	0·026
BS_2	0·04	0·32	0·04	0·4	0·182	0·820	0·102
BS_3	0·0	0·06	0·34	0·4	0·0	0·154	0·872
	0·22	0·39	0·39	1·0			

Figure 14. *Posterior and Pre-Posterior Analysis of Expected Profits in the Japanese Market in the Light of Given Test Market Shares* (£m.)

	Posterior			*Preposterior*		
Observed Test Brand Share	BS_1 (i)	BS_2 (ii)	BS_3 (iii)	Implicit Profit (iv)	Probability (v)	Profit Consequences (iv) × (v)
TS_1	6·544	0·364	0·0	6·908	0·22	1·520
TS_2	0·208	1·640	–0·662	1·186	0·39	0·463
TS_3	0·208	0·204	–3·749	–3·337	0·39	0·0

Expected value of introducing in Japanese after test-market research ....£1·983m.

Figure 15. *Canadian Market*

Broadscale Brand Shares	*Probability of Joint Occurrence of Test and Broadscale Brand Shares*			*Marginal Total*	*Posterior Probabilities Taking Cognizance of Test Data*		
	TS_1	TS_2	TS_3		TS_1	TS_2	TS_3
BS_1	0·27	0·015	0·015	0·3	0·844	0·034	0·064
BS_2	0·05	0·4	0·05	0·5	0·156	0·899	0·213
BS_3	0·0	0·03	0·17	0·2	0·0	0·067	0·723
	0·32	0·445	0·235	1·0			

Figure 16. *Posterior and Pre-Posterior analysis of Expected Profits in Canadian Market in the Light of Given Test Market Shares (£m.)*

	Posterior			*Preposterior*		
Observed Test Brand Share	BS_1 (i)	BS_2 (ii)	BS_3 (iii)	Implicit Profit (iv)	Probability (v)	Profit Consequences (iv) × (v)
TS_1	25·32	1·872	0·0	27·192	0·320	8·701
TS_2	1·02	10·788	–1·541	10·267	0·445	4·569
TS_3	1·92	2·556	–16·629	–12·153	0·235	0·0

Expected value of introducing in Canadian after test market research ....£13·270m.

Figure 17. *Network of Discount Expected Value of Introducing in the Canadian Market with and Without Test Marketing*

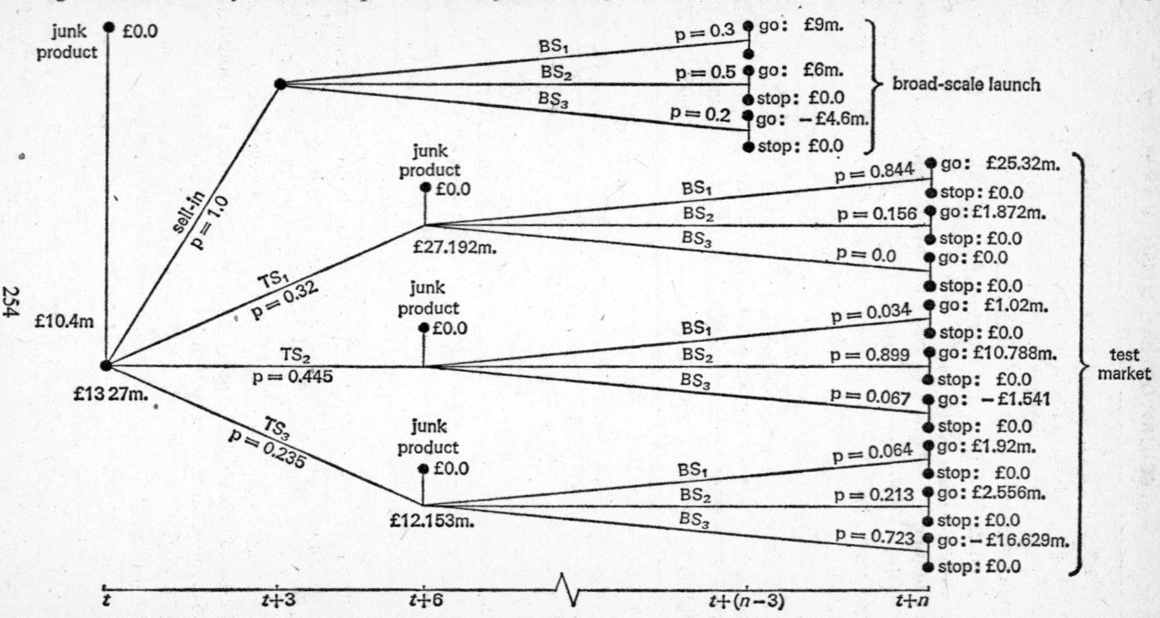

For Canada, Figure 16 shows an increased value of introduction with test-market information of £13·27m.−£10·4m. = £2·87m. Once again, if the probabilities provided in Figure 12 were credible for a test marketing cost of £0·25m., a test would clearly be optimal.

Figures 14 and 16 are derived by posterior and preposterior analyses, which can be illustrated as follows in Figure 14:
Column (i): 6·544m. = 0·818 (posterior probability Figure 13, TS_1/BS_1)$\times$£8m. (implicit profit of BS_1 Figure 11); likewise, columns (ii) and (iii). Column (iv) is the sum of the implicit profits of any test market share being achieved. This is multiplied by the probabilities (derived from Figure 13 as $\Sigma\, TS_1$ etc.) to provide the profit consequences which, when the optimal course in each case is summed, indicates the expected value of introducing our market offering into Japan with test-market information. For each test-market share observed, the TMA group will have the option of continuing or ignoring the market. If TS_3 materializes the loss of £3·337m. will be avoided in favour of not entering the Japanese market. With TS_1 and TS_2, however, broad-scale marketing would go ahead.

Figure 17 uses a decision tree to describe the alternative courses of action, and their profit consequences, for Canada. This is a more comprehensible version of the data given in Figures 15 and 16.

MISSION MANAGEMENT

The complexity of TMA will have by now become fully apparent. The management planning, coordination and control of all such analysis, and its accompanying activities in R & D and market testing, themselves demand formalization. Networking and critical path method (CPM) have been developed to accomplish this task, most noticeably originating in the missile and space programmes of the 1950s and 1960s. They are now widely employed in all manner of project management.

They are essentially an extension of the nineteenth-century Gantt charting procedures, which identified all the relevant activities and demonstrated how their patterns of fulfilment overlapped. Networking does the same thing, but, by computerizing the entire procedure, business is able to have fingertip control over movement towards any particular programmed goal. Equally important, however, is the opportunity which emerges for finding the critical path through a project and any *slack time* which may be present on non-critical paths. This normally means that resources can be redeployed away from slack areas into areas of criticality, often reducing the time-scale for project completion. This is, of course, assuming that *early* completion is desirable.

There are two main elements in a network: an *activity* or time consuming task, and an *event* or the accomplishment which occurs at a point in time. In Figure 18, events are shown as boxes and activities are represented by lines. Events have been given roman numerals. The EED value associated with each event is the *earliest expected date* by which that event can be completed, i.e. the sum of the maximum times taken to accomplish all previous events. Each activity has its own time estimate (t_e) which is computed as follows:

$$t_e = \left[\frac{\begin{pmatrix} \text{most} \\ \text{optimistic} \\ \text{time} \end{pmatrix} + \begin{pmatrix} \text{4 times} \\ \text{most likely} \\ \text{time} \end{pmatrix} + \begin{pmatrix} \text{most} \\ \text{pessimistic} \\ \text{time} \end{pmatrix}}{6} \right]$$

The critical path through this simple network can be seen to be events 0–I–IV–VII with a minimum time for completion of seventy-two months. Slack time exists within the network on all non-critical paths between events. Slack time can be defined as the difference between the earliest achievable time for an event and the latest permissible time. Event II must not be completed later than 40–28, i.e., twelve months after event 0. It can be achieved within eight; hence slack time is four months. Managements may be able to reallocate some of their resources from achieving event II to event I (unlikely in this instance) to reduce the EED time of forty months associated with event IV. Without realloca-

Figure 18. *Network with Time Estimates for Technological Mission*

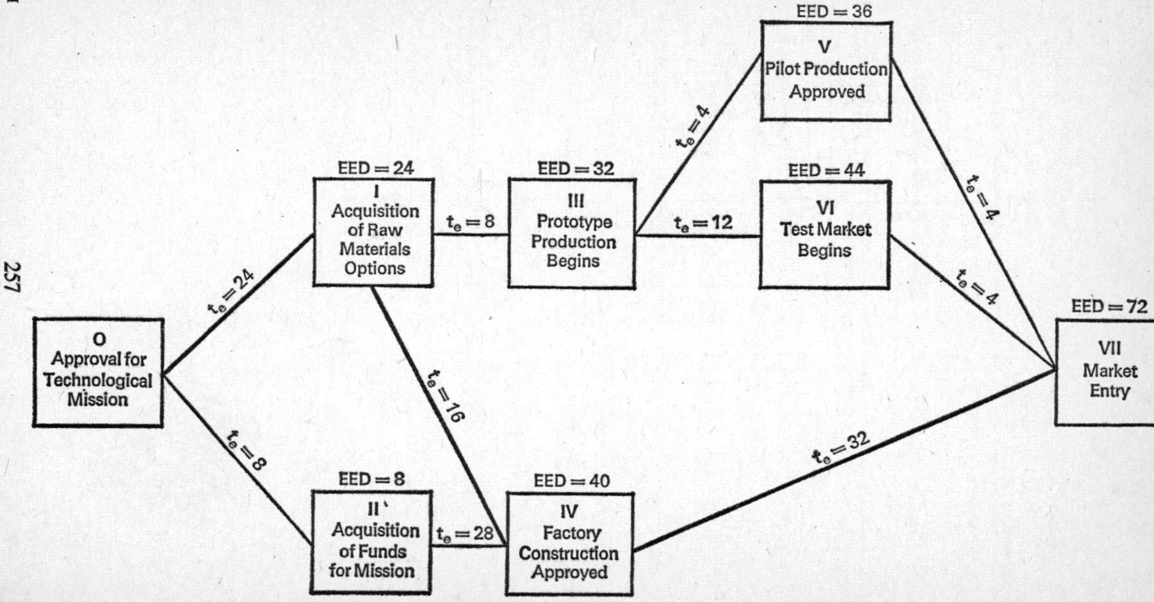

Figure 19. *Summary Network for New Product Launch*

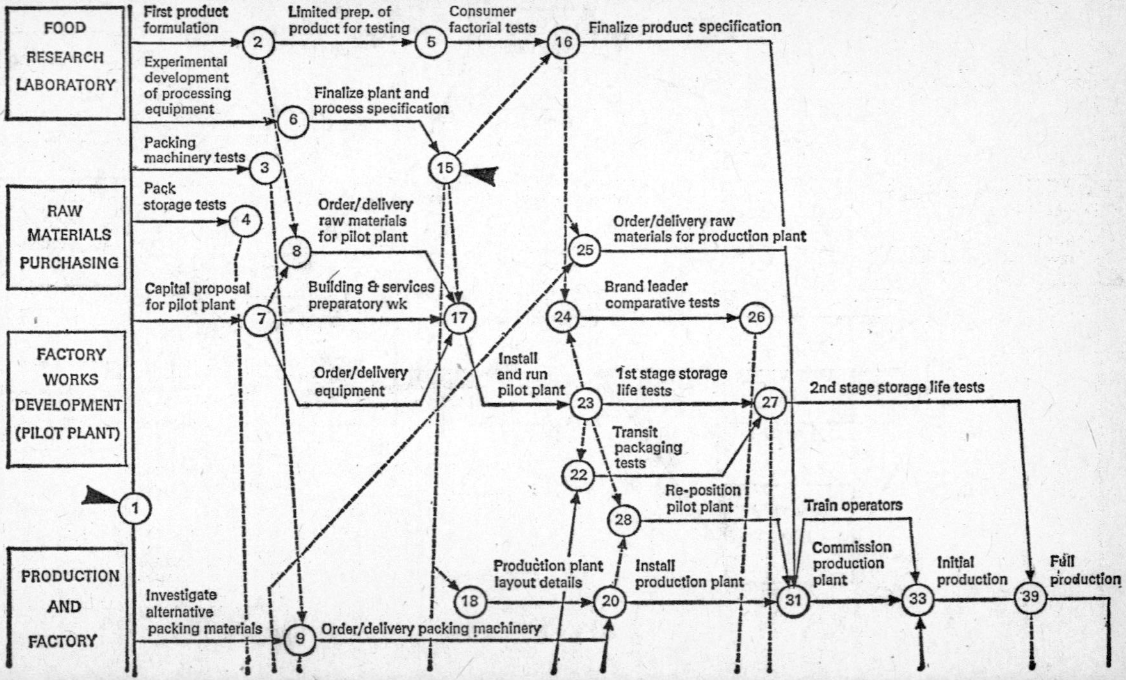

FOOD RESEARCH LABORATORY

RAW MATERIALS PURCHASING

FACTORY WORKS DEVELOPMENT (PILOT PLANT)

PRODUCTION AND FACTORY

First product formulation

Limited prep. of product for testing

Consumer factorial tests

Finalize product specification

Experimental development of processing equipment

Finalize plant and process specification

Packing machinery tests

Pack storage tests

Order/delivery raw materials for pilot plant

Order/delivery raw materials for production plant

Capital proposal for pilot plant

Building & services preparatory wk

Brand leader comparative tests

Order/delivery equipment

Install and run pilot plant

1st stage storage life tests

2nd stage storage life tests

Transit packaging tests

Re-position pilot plant

Train operators

Commission production plant

Production plant layout details

Install production plant

Initial production

Full production

Investigate alternative packing materials

Order/delivery packing machinery

1 2 3 4 5 6 7 8 9 15 16 17 18 20 22 23 24 25 26 27 28 31 33 39

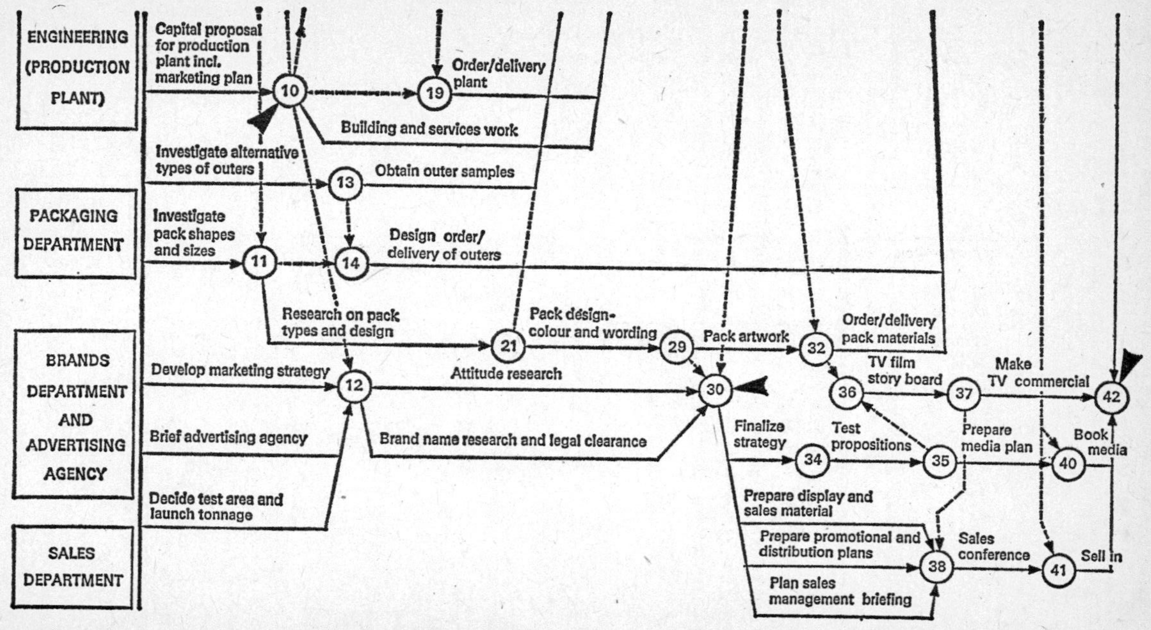
ENGINEERING (PRODUCTION PLANT)
PACKAGING DEPARTMENT
BRANDS DEPARTMENT AND ADVERTISING AGENCY
SALES DEPARTMENT
Capital proposal for production plant incl. marketing plan
Investigate alternative types of outers
Investigate pack shapes and sizes
Develop marketing strategy
Brief advertising agency
Decide test area and launch tonnage
10
19
Order/delivery plant
Building and services work
13
Obtain outer samples
11
14
Design order/ delivery of outers
Research on pack types and design
12
21
Pack design- colour and wording
Attitude research
Brand name research and legal clearance
29
Pack artwork
30
32
Order/delivery pack materials
Finalize strategy
34
Test propositions
35
36
TV film story board
37
Make TV commercial
Prepare media plan
40
Book media
42
Prepare display and sales material
Prepare promotional and distribution plans
Plan sales management briefing
38
Sales conference
41
Sell in

tion within this particular mission, savings may be made by reallocation to other work on different missions of the personnel engaged in the activity leading to the completion of event I and the onward activity to event IV.

Once the network and critical path have been identified business management knows where to focus its attention to ensure that the overall project duration of seventy-two months is not extended. Manpower and financial resource allocations can be harnessed to this system and week-by-week control of the project obtained, providing opportunities to update time estimates and create fresh critical paths and slack time.

An advanced application of critical path method of networking has been provided by McLaren and Buesnel of Unilever. It appears as Figure 19 and demonstrates the great complexity of such management problems even within one sector of TMA.

Concluding Remarks

Technological forecasting holds out exciting prospects both for society and the individual business. It is an area of management technique about which a great deal more will be heard as technology plays an increasingly important part in our lives. Almost inevitably we can anticipate that its deployment will frequently be confused by the terminology which is rapidly emerging amongst its practitioners.

This book has been written at the very onset of the development of technological forecasting in an attempt to forestall some of that potential confusion. Morphological analysis, relevance trees, the Delphi method of time-scaling, scenario writing and the rest are all valuable techniques in the hands of wise men but dangerous in the hands of the ignorant or foolish. All techniques are the same.

All the evidence from survey work conducted in continental Europe and in Britain at the end of the sixties showed that firms have not yet initiated the full panoply of technological forecasting procedures in their businesses. Those who were using some techniques were doing so with due caution and seeking to place them logically and effectively within the overall corporate planning context. This is an absolutely correct approach, and we can only hope it will be widely emulated by other firms who begin to embrace technological forecasting. Technological forecasting is no panacea for research and development planning or for corporate profitability; it is just one more aid to be kept in its correct perspective. An important part of that perspective is the overall theme of technological mission analysis which we have just considered. A yet wider setting is, of course, corporate planning. The widest, and perhaps ultimately the most important setting, is society at large. Hence it is a tool both for organizations such as the United Nations, OECD and national governments as well as for the individual business.

I trust that in these pages I have been able to convey some of the excitement I feel about the potential of technological forecasting to assist in the more purposeful building of our society.

Glossary of Major Terms used in Technological Forecasting

The basic terminology has been well-defined although some writers show wider or narrower views on some definitions.

The following terms are adopted because:

(*a*) they are simple and comprehensive; and

(*b*) they correspond to the actual pattern in existence at the operational level.

Anticipation:	A logically constructed model of a possible future, on a confidence level as yet undefined.
Delphi Method:	Empirical derivation of a consensus of expert opinion concerning the time-scale over which posited technological futures may be realized.
Envelope Curve:	That curve which envelops a series of S-curves through time to provide exponential growth in functional capabilities from a series of more advanced technologies.
Exploratory Technological Forecasting:	Starts from today's assured basis of knowledge and is oriented towards the future, i.e., includes extrapolation and morphological analysis.
Extrapolation:	The projection of existing trends in, for example, functional capabilities for given technological parameters, in order to establish the direction in which current developments will carry capability other things being equal, and the time-scale for such development.
Forecast:	A probabilistic statement, on a relatively high confidence level, about the future.
Functional Research:	Research connected with present activities and the organic extension of such research.
Fundamental Research:	Research on the fundamentals of science and technology.

Fundamental Scientific Research: Research which broadly pertains to the level of scientific resources in the technology transfer space.

Fundamental Technological Research: Research connected with present activities and the organic extension of such research.

Impact Analysis: Analysis of the likely impact on other sectors of new technological developments.

Models: Representations of processes describing in simplified form some aspects of the real world.

Morphological Analysis: Analysis of the structure of technological systems into basic parameters and the exhaustive identification of the alternative ways of fulfilling those parameters. The matrix so prepared affords the opportunity to explore new combinations.

Non-Functional Research: Research concerned with future new activities.

Normative Technological Forecasting: (1) Forecasting that first assesses future goals, needs, desires, missions, etc., and then works backwards to the present. (2) A coherent examinaation of future needs in a future society, helping to define socio-economic objectives, and then purely technical research objectives, and the best way to achieve them.

Normex Reconciliation: A forecasting procedure which matches normative scenarios concerning future market offerings with the extrapolation of technical capabilities.

Prediction: A non-probabilistic (or apodictic) statement, on an absolute confidence level, about the future. As such, it is the objective of the forecasting exercise.

Projection: A datum in a forecasting exercise.

Relevance Trees: Diagrammatic representation of the sequential paths through which development must go in order to achieve a technological mission. Allied to networking and critical-path method, it acts as a powerful tool for planning, coordinating and reviewing progress towards a goal.

S-Curve: The most common shape taken by the curve of functional capability delivered by any specific technology over time, i.e., a slow initial development followed by rapidly improving capability and then exhaustion.

Scenario: Postulated future technological capability, serving as a creative input to technological forecasts. The probability of its attainment and the time-scale for development are not known.

Simulation: The operation of a model by manipulation with a computer, human player, or both.

Social Engineering: An invention with considerable potential impact on the technology transfer levels of social systems and society.

Social Technology: Technology with substantial implications for society. Frequently it is based on social invention: an invention with considerable potential impact on the technology transfer levels of social systems and society.

SOON Chart: Analytical approach to the sequence of opportunities and negatives in achieving a technological future. *See* Relevance Trees.

Technological Change: Any change in the technology transfer space effected by technology transfer.

Technological Forecasting: (1) The probabilistic assessment on a relatively high confidence level of future technology transfer. (2) Forecasting innovations on a time-scale, which should be restricted to the anticipation of such changes in technology as will have an impact on the firm's future, and should not only outline some possible new fields for technical discoveries, but should also provide ways of achieving these innovations using the optimum technical paths.

(3) The forecasting of technological change.
(4) The prediction of the invention, characteristics, dimensions, or performance of a machine serving some useful purpose.
(5) The description or prediction of a foreseeable invention, specific scientific refinement, or likely scientific discovery that promises to serve some useful function.

Technological Mapping: Conscious effort to piece together information which will indicate the pattern of competitive technological effort in order to make a company's own management better able to forecast the time dimension and direction of competitive activities.

Technological Mission: The technological goal set by a company after evaluating the alternative strategies open to it. Technological mission analysis is the detailed examination of the processes by which a given goal can be obtained in order that the optimum route for technological development can be followed and technology transfer most effectively made. This represents an extension of the more traditional concept of venture analysis to include all evaluation and development stages once a technological forecast has been adopted as a corporate mission.

Technological Planning: The development of an intellectual concept concerned with the active implementation of technological transfer – both vertical and horizontal.

Technology: The broad area of purposeful application of the physical, life, and behavioural sciences. It comprises the entire notion of technics, as well as the medical, agricultural, management and other field, as with their total hardware and software contents.

References Cited in the Text and Suggestions for Further Reading

The books and papers referred to here against each section of this volume are the major sources of the ideas which are presented. Whilst I have done my very best to encapsulate accurately the opinions of others, there can be no substitute for looking at their work at first hand. Accordingly, I hope that these listings may act as a starting point bibliography for any further examination of TF which readers may be moved to undertake.

Introduction *Technological Myopia*

Arla Chemical Company, (1962), Case Study from IMEDE, Lausanne. The name of the company has been changed to conceal its identity; the facts have not. Written by Learned, E. P., and Aguilar, F. J.

Beer, S., Wills, G. S. C., (1969), 'Government Money for the Inventor', *Management Decision*, 3, 2, Summer.

British Industry Week (1968), Special Survey: *Technology in Britain; a Test of Strength*, 233, 3 May.

Burns, T., (1969), *Models, Images and Myths*, Ch. 1, in Gruber, W. H., and Marquis, D. G. (see below).

Catherwood, F., (1969), 'The Planning Dialogue', *National Westminster Bank Review*, May, pp. 2–9.

Galbraith, J. K., (1958), *The Affluent Society*, Hamish Hamilton, Penguin Books.

Galbraith, J. K., (1967), *The New Industrial State*, Hamish Hamilton.

Gruber, W. H., Marquis, D. G. (eds.), (1969), *Factors in the Transfer of Technology*, M.I.T. Press, Cambridge, Mass.

Guardian, The (1967), 'Think Tanks', 26 September.

Hall, P. D., (1968), 'Technological Forecasting for Computer Systems', Ch. 12 of Wills, G. S. C., Ashton, D., Taylor, B. (1969) (see below).

Hayhurst, R., Wills, G. S. C., (1971), *Organizational Design for Marketing Futures*, George Allen & Unwin. The implications of the computer for entrepreneurship are more fully explored in Part I of this book, entitled 'Marketing Futures'.

Hetrick, J. C., (1968), 'The Impact of Technological Forecasting on Long Range Planning', Ch. 3 of Wills, G. S. C., Ashton, D., Taylor, B., (1969) (see below).

Leavitt, T., (1960), 'Marketing Myopia', *Harvard Business Review*, July/August.

Peters, Lenrie, (1964), poem in Mphahlele, E. (ed)., (1967), *African Writing Today*, Penguin Books, p. 243.

Reynolds, W. B., (1961), 'Research and the Marketing Concept', Part 2, pp. 14–21, of *Marketing Innovations*, Proceedings of the 8th Biennial Marketing Institute, American Marketing Association, Minnesota Chapter, November.

Schon, D. A., (1967), *Technology and Change: the New Heraclitus*, Pergamon Press.

Servan-Schreiber, J. J., (1968), *Le Défi Americain* (*The American Challenge*), Hamish Hamilton, Penguin Books.

Stravinsky, I., (1945), *Musical Poetics*, La Flute de Pau, pp. 89/99.

Wills, G. S. C., (1968), 'Technological Forecasting; the Art and its Management', *Commentary*, 10, 2, pp. 87–101, April.

Wills, G. S. C., Ashton, D., Taylor, B. (eds.), (1969), *Technological Forecasting and Corporate Strategy*, Bradford University Press/ Crosby Lockwood; and American Elsevier.

Wills, G. S. C., (1969), 'Technological Myopia', *Management Decision*, 3, 4, pp. 48–53. (This Introduction is largely based on this article and is used with permission.)

Wills, G. S. C., (1970), The Development and Deployment of Technological Forecasts, *Journal of Long Range Planning*, 2, 3, March, pp. 44–55.

PART ONE THE MANAGERIAL IMPLICATIONS OF TECHNOLOGICAL FORECASTING

Chapter 1.1 *The Technological Setting*

Arnfield, R. (ed.), (1969), *Technological Forecasting* Edinburgh University Press.

Beattie, C. McG., and Fraser, J. C., (1968), 'The Impact of Technological Forecasting on Marketing', *University of Bradford Management Centre Project Report*, July.

Blau, P. M., (1964), *Exchange and Power in Social Life*, Wiley.

Bright, J. R., (1963), 'Opportunity and threat in Technological Change', *Harvard Business Review*, 41, 6.

Cazes, B., (1969), 'The Proper Use of Long Term Forecasts', *Sixth Congress of International Council of Industrial Design*, London September.

de Jouvenal, in Young, M., (1969); (see below).

Horowitz, I. L. (ed.), (1964), *The New Sociology*, Oxford University Press.

Jantch, E., (1967), *Technological Forecasting in Perspective*, OECD.

Kahn, H., and Wiener, A. J., (1967), *The Year 2000*, Macmillan.

Ministry of Technology, (1967), *Papers Presented to TUC Productivity Conference*, November.

Ozbekhan, H., (1969), 'The Future as an Ethical Concept'. *Sixth Congress of International Council of Industrial Design*, London, September.

Pickering, W. H., (1908), Speech at Harvard University, June.

Quinn, B., (1967), 'Technological Forecasting', *Harvard Business Review*, 45, 2, pp. 84–95.

Young, M. (ed.), (1969), *Forecasting and the Social Sciences*, SSRC.

Chapter 1.2 *The Environmental Impact*

Arnfield, R. (ed.), (1969), *Technological Forecasting*, Edinburgh University Press.

Benn, A. Wedgewood, Minister of Technology, (1968), *European Technological Collaboration*, Speech to Council of Europe, 26 January.

Bredemeir, H. C., and Stephenson, R. M., (1962), *The Analysis of Societal Systems*, Holt, Rinehart, and Winston.

Fisher, C. J. C., (1967), *Social Theory and Social Structure*, Collier Macmillan/Free Press of Glencoe.

Galbraith, J. K., (1967), *The New Industrial Estate*, Hamish Hamilton.

Halsbury, Earl of, (1957), *Integration of Social With Technological Change*, Institute of Personal Management, Occasional Paper No. 11.

Johnson, H. M., (1963), *Sociology*, Routledge and Kegan Paul.

Merton, R. K., (1949), *Social Theory and Social Structure*, Collier Macmillan/Free Press of Glencoe.

Parsons, T., (1951), *The Social System*, Routledge and Kegan Paul.

Report of the President's Commission, (1960), *Goals for Americans*, Prentice Hall, pp. 3–20.

Schon, D., (1969), 'Design in the Light of the Year 2000', *Sixth Congress of International Council of Industrial Design*.

Simon, H. A., and March, J. G., (1958), *Organisations*, Wiley, p. 65.
Sulc, O., (1968), *Forecasting the Interactions Between Technological and Social Change*, Manchester Business School (mimeograph), July.
Wilson, R. M. S. (1969), 'The Management of Corporate Intelligence', *Factory Management*, 38, 10, October.

Chapter 1.3 *The Corporate Impact*

Adler, L., (1966), 'Timelag in New Product Development', *Journal of Marketing*, 30, 1, pp. 17–21.
Ansoff, H. I., and Stewart, J. M., (1967), 'Strategies for Technology based Business', *Harvard Business Review*, November/December, pp. 71–83.
Anthony, R. N., (1965), *Planning and Control Systems*, Harvard University Press.
Burns, T., and Stalker, G. M., (1961), *The Management of Innovation*, Tavistock Publications.
Hirsch, S., (1967), *Location of Industry and International Competitiveness* Oxford University Press.
Jantch, E., (1967), *Technological Forecasting in Perspective*, OECD.
Nielsen, A. C., (1966), 'Marketing Laboratory', *Nielsen Researcher*, 7, January/February, pp. 3–5.
Quinn, J. B., (1967), 'Technological Forecasting', *Harvard Business Review*, March/April, pp. 84–95.
Quinn, J. B., (1968), 'Technological Strategies for Industrial Companies', *Management Decision*, 2, 3, pp. 182–8.
Wills, G. S. C., (1970), 'The Preparation and Deployment of Technological Forecasts', *Journal of Long Range Planning*, 2, 3, pp. 44–55.
Zuckerman, Sir S., (1967), *The Image of Technology*, 4th Maurice Lubbock Memorial Lecture, Oxford University Press, pp. 17–18.

Chapter 1.4 *The Functional Impact with Business*

Ansoff, A. I., and Stewart, J. M., (1967), 'Strategies for a Technology-based Business', *Harvard Business Review*, 45, 6, pp. 71–83.
Beattie, C. McG., and Fraser, J. C., (1968), *The Impact of Technological Forecasting on Marketing*, University of Bradford Management Centre Project Report, July.
Blackman, A. W., (1969), 'The Use of Technological Forecasting in an R & D Planning and Budgeting System', *Advanced TF Conference, Lake Placid Club*, NY, September.

Bright, J. R. (ed.), (1968), *Technological Forecasting for Industry and Government*, Prentice Hall.
Brown, R. V., (1969), in Arnfield, R., *Technological Forecasting*, Edinburgh University Press.
Carter, C. F., and Williams, B. R., (1957) *Industry and Technical Progress*, Oxford University Press.
Carter, C. F., and Williams, B. R., (1958) *Investment in Innovation*, Oxford University Press.
Clarke, Sir Richard, (1968), 'The Contribution of Technology to Industry and The Economy', *National Conference, Institute of Marketing*.
Goodwin, G. D., (1969), in Arnfield, R., *Technological Forecasting*, Edinburgh University Press.
Kahn, H., and Weiner, A. J., (1967), *The Year 2000*, Macmillan.
Kramish, A., (1969), 'Thinking about the Thinkable: toward a Global Design for the Future', *Sixth Congress of International Council of Industrial Design*.

PART TWO TF THE FORECASTING ART

Chapter 2.1 *Extrapolative Approaches*

Ayres, R. U., (1969), *Technological Forecasting and Long Range Planning*, McGraw-Hill, pp. 94–142.
Bright, J. R. (ed), (1968), *Technological Forecasting for Industry and Government*, Prentice Hall, Part 2, pp. 57–109 and 144–80. These pages include the contributions cited in the text by Lenz, R. C. jnr., Ayres, R. V., and Floyd, A. L.
Design and Innovation Group Symposium, (1969), University of Aston in Birmingham, September. The proceedings of this conference include papers by Gregory, S. A., and Brookes, L. G., which are well worth the reader's attention. Mimeographs.
Hobbs, J. A., (1969), 'Trend Projection', in Arnfield, R. (ed)., *Technological Forecasting*, Edinburgh University Press, pp. 231–40.
Jantch, E., (1967), *Technological Forecasting in Perspective*, OECD, pp. 143–73.

Chapter 2.2 *Morphological Analysis*

Ayres, R. U., (1969), *Technological Forecasting and Long Range Planning*, McGraw-Hill, pp. 72–86.

Bridgewater, A. V. (1969), 'Morphological Methods: Principles and Practice', in Arnfield, R., *Technological Forecasting*, Edinburgh University Press, pp. 211–30.

Design and Innovation Group Symposium, (1969), University of Aston in Birmingham Session 1, *Morphological Methods:* two important papers by Gregory, S. A., and Wills, R. J./Hawthorne, E. P.

Gardner, M., (1958), *Logic Machines and Diagrams*, McGraw-Hill.

Garrett, T., and others, (1969), 'Illustrations of TF for R & D Evaluations', in Wills, G. S. C., Ashton, D., and Taylor, B. (eds.), *Technological Forecasting and Corporate Strategy*, Crosby Lockwood for Bradford University Press.

Norris, K. W., (1963), 'The Morphological Approach to Engineering Design', in Jones, J. C., and Thornley, D. G. (eds.), *Conference on Design Methods*, Pergamon Press.

Poyner, B., (1969), *The Geometry of Office Floor Space which permits sub-division into Small Tenancies*, Research Paper No. 2, School of Architecture, University of Aston-in-Birmingham.

Reuleaux, F., (1963), *The Kinematics of Machinery*, (trans. 1876), Dover Books.

Rothman, L. J., and Tate, B., (1964), 'Research Techniques for Maniple Marketing', in *Research in Marketing*, Market Research Society.

Wankel, F., (1957), *Rotary Piston Engines*, Iliffe Books.

Zwickey, F., (1962), *Monographs on Morphological Research*, Society for Morphological Research, Pasadena, California.

Chapter 2.3 *Normative Approaches and Normex Reconciliation*

Arnfield, R. (ed.), (1969), *Technological Forecasting*, Edinburgh University Press, pp. 170–77; 197–210; 267–306.

Blackman, A. W., (1969), 'The Use of Technological Forecasting in an R & D Planning and Budgeting System', Appendix 3; a mimeograph presented to *Advanced TF Conference*, Lake Placid Club, NY, September.

Brech, R., (1963), *Britain 1984: Unilever's Forecast*, Darton, Longman, and Todd.

Gabor, D., (1964), *Inventing the Future*, Penguin Books.

Hayhurst, R., and Wills, G. S. C., (1971), *Organizational Design for Marketing Futures*, George Allen & Unwin.

Jantch, E., (1967), *Technological Forecasting in Perspective*, OECD, pp. 180–81; 211–33.

Kahn, H., and Wiener, A. J., (1967), *The Year 2000*, Macmillan.

Ozbekhan, H., (1967), Automation, *Science Journal*, 3, pp. 67–72.
Wills, G. S. C., Ashton, D., and Taylor, B., (eds.), (1969) *Technological Forecasting and Corporate Strategy*, pp. 190–206; 220–51.

Chapter 2.4 *Delphi Method of Time-scaling*

Dalkey, N., and Helmer, O., (1963), 'An Experimental Application of the Delphi Method to the Use of Experts', *Management Science*, 9, 3, pp. 458–67.
Gordon, T. J., and Hayward, H., (1968), Initial Experiments with the Cross-Impact Matrix Method of Forecasting, *Futures*, 1, 2, pp. 100–116.
Gordon, T. J., and Helmer, O., (1964), *Report on a Long Range Forecasting Study*, Rand Corporation, p. 2982.
Gordon, T. J. (1968), 'New Approaches to Delphi', pp. 134–43; Helmer, O., (1968), 'Analysis of the Future: The Delphi Method', pp. 116–143; both in Bright, J. (ed.), *Technological Forecasting for Industry and Government*, Prentice Hall.
Hall, P. D., (1969), 'Computer Systems', in Wills, G. S. C., Ashton, D., and Taylor, B., *Technological Forecasting and Corporate Strategy*, Crosby Lockwood for Bradford University Press, pp. 191–206.
Mandanis, G. P., (1969), The Future of The Delphi Technique, in Arnfield, R. (ed.), *Technological Forecasting*, Edinburgh University Press, pp. 159–69.
North, H. Q., and Pyke, D. L., (1968), 'Corporate Experience with Delphi', presented at *2nd Annual Technology and Management Conference*, Washington, D. C., March.
Ozbekhan, H., (1967), 'Automation', *Science Journal*. 3, pp. 67–72.
Parker, E. T., (1969), 'The Blue Sky Problem', presented at *University of Aston Design and Innovation Group Symposium*, September.
Payne, S., (1951), *The Art of Asking Questions*, Princeton University Press.

Chapter 2.5 *Technological Mission Analysis*

Anderson, Sigurd L., (1957), 'Venture Analysis – A Flexible Planning Tool', *Chemical Engineering Progress*, 57, May–June.
Blackman, A. W., (1969), 'The Use of Technological Forecasting in an R & D Planning and Budgeting System', *Advanced TF Conference, Lake Placid Club*, N.Y., September.
Christian, Richard C., (1959), 'A Checklist for New Industrial Products', *Journal of Marketing*, 24, July.

Christopher, Martin, (1969), *Venture Analysis*, University of Bradford Management Centre mimeograph, September.

Hanan, M., (1969), 'Corporate Growth Through Venture Management', *Harvard Business Review*, 47, January-February.

Harder, V. E., and Lindell, F. R., (1966), 'Using PERT in Marketing Research', *Business Horizons*, 9, Summer, pp. 97–102.

Johnson, S .C., and Jones, C. (1957), 'How to Organize for New Products', *Harvard Business Review*, 35, May–June.

O'Meara, John T., Jnr., (1961), 'Selecting Profitable Products', *Harvard Business Review*, 39, January–February.

Pessemier, Edgar, A., (1966), *New Product Decisions – an Analytical Approach*, McGraw-Hill.

Peterson, Russell W., (1967), 'New Venture Management in a Large Corporation', *Harvard Business Review*, 45, May–June.

Richman, Barry M., (1962), 'A Rating Scale for Product Innovation', *Business Horizons*, 5, Summer.

Scheuble, Philip A., Jnr., (1964), 'ROI for New-Product Policy', *Harvard Business Review*, 42, November–December.

Sponsor Groups – S. C. Johnson Effective Screening of New Products, *Product Engineering*, 28, October 1957.

Wills, G. S. C., 'Cost-Benefit Analysis of a Test Market', *Management Decision* 1, 4, pp. 17–21; reprinted in Seibert, J., and Wills, G. S. C., (eds.) (1970), *Marketing Research*, Penguin Books, pp. 88–103.

Winer, L., 'A Profit-orientated Decision System', *Journal of Marketing*, 30, 1, pp. 38–44.

Index

PROGRESS OF MANAGEMENT RESEARCH

Edited by Nigel Farrow

'Management research', writes Nigel Farrow, 'is science's Oliver Twist: a delicate and neglected infant of obscure parentage, it has been suddenly claimed by various competing godfathers for reasons ranging from disinterested charity to commercial exploitation.'

This volume in the Pelican Library of Business and Management contains ten articles which originally appeared as a series in *Business Management*. It is a sign of the fluid state of management studies that the contributors include professors of marketing, business administration, industrial psychology, operational research, and industrial and management engineering, as well as economists and consultants. Covering the functional areas of production, personnel, finance, and marketing, they indicate the debt owed by business research to economics, mathematics, psychology, and sociology.

It remains a question whether management research does better to be wide, general, and abstract (a pursuit for academic cloisters); or specific, local, and concrete (an exercise for the oil-grimed shop). But in any case the recent research outlined in these essays is constructive and practical and never loses sight of the manager on the spot.

AN INSIGHT INTO MANAGEMENT ACCOUNTING

John Sizer

Management accountancy is the key to modern business strategy and technique. No department specialist – and certainly no general executive – can cope without an insight into its principles and practices. This book is designed to give just such an insight: to enable every businessman to understand the finances and internal costing of his company – and to keep its accountants on their toes.

John Sizer, who has practical experience of several industries, is now Senior Tutor in Accounting at the London Graduate School of Business Studies. In this book he discusses such subjects as stewardship, cost accounting, the measurement and control of profitability, long-range planning, capital investment appraisal, budgetary control and marginal costing. Other chapters cover the accountant's contribution to the pricing decision, and company taxation. But whilst he provides an admirably succinct description of modern accounting techniques, John Sizer's main achievement has been to relate them all to essentially functional and practical situations within the firm. This is a no-nonsense manual which firmly removes from accounting all the mystique which overawes managers and often merely masks muddle.

THE GENESIS OF MODERN MANAGEMENT

Sidney Pollard

In the cut-throat competition of the Industrial Revolution many capitalists saw the need for 'managers' – employees paid to plan profits, keep accounts and mould men to the new machines. Living down their rascally reputation for dishonesty, these supervisors eventually came to constitute a new class, as essential and practical a part of the impetus to industrialism as the working class itself.

This volume in the Pelican Library of Business and Management is a unique study of this early managerial revolution, of the problems which confronted the first generation of managers and of the complex interactions of mass-production technology and human organization. An important contribution to developmental economics, it is also a convincing analysis of the historical factors which continue to condition the present-day practices of British management.

MANAGEMENT THINKERS

Edited by A. Tillett, T. Kempner and G. Wills

'British management was backward by the end of the nineteenth century and, with notable exceptions, has never caught up.'

'The Ford Co. from October 1912 to October 1913 hired 54,000 men to maintain an average working force of 13,000. This was a labour turnover of 416 per cent for the year.'

With observations like these this book vividly recalls and presents the problems of industry from the industrial revolution to the present day.

Management Thinkers contains readings from the pioneers of management thought – Frederick Taylor, Henri Fayol, Seebohm Rowntree, Mary Parker Follett, Elton Mayo and C. I. Barnard.

The problems of industry are the problems of today. This is a history of how these problems have been tackled in the past, together with a careful assessment of the contribution the management thinkers have made to the world we live in now.